IFMBE Proceedings

AF822805

132

The IFMBE Proceedings Book Series is an official publication of *the International Federation for Medical and Biological Engineering* (IFMBE). The series gathers the proceedings of various international conferences, which are either organized or endorsed by the Federation. Books published in this series report on cutting-edge findings and provide an informative survey on the most challenging topics and advances in the fields of medicine, biology, clinical engineering, and biophysics.

The series aims at disseminating high quality scientific information, encouraging both basic and applied research, and promoting world-wide collaboration between researchers and practitioners in the field of Medical and Biological Engineering.

Topics include, but are not limited to:

- Diagnostic Imaging, Image Processing, Biomedical Signal Processing
- Modeling and Simulation, Biomechanics
- Biomaterials, Cellular and Tissue Engineering
- Information and Communication in Medicine, Telemedicine and e-Health
- Instrumentation and Clinical Engineering
- Surgery, Minimal Invasive Interventions, Endoscopy and Image Guided Therapy
- Audiology, Ophthalmology, Emergency and Dental Medicine Applications
- Radiology, Radiation Oncology and Biological Effects of Radiation
- Drug Delivery and Pharmaceutical Engineering
- Neuroengineering, and Artificial Intelligence in Healthcare

IFMBE proceedings are indexed by SCOPUS, EI Compendex, Japanese Science and Technology Agency (JST), SCImago. They are also submitted for consideration by WoS.

Proposals can be submitted by contacting the Springer responsible editor shown on the series webpage (see "Contacts"), or by getting in touch with the series editor Ratko Magjarevic.

Chwee Teck Lim · Hwa Liang Leo ·
Changsheng Wu · Andrew Holle ·
Wei Seong Toh · Raye Chen-Hua Yeow
Editors

18th International Conference on Biomedical Engineering

Selected Contributions to ICBME 2024
December 9–12, 2024
Singapore

Springer

Editors
Chwee Teck Lim
Department of Biomedical Engineering
National University of Singapore
Singapore, Singapore

Hwa Liang Leo
Department of Biomedical Engineering
National University of Singapore
Singapore, Singapore

Changsheng Wu
Department of Materials Science
and Engineering
National University of Singapore
Singapore, Singapore

Andrew Holle
Department of Biomedical Engineering
National University of Singapore
Singapore, Singapore

Wei Seong Toh
Department of Biomedical Engineering
National University of Singapore
Singapore, Singapore

Raye Chen-Hua Yeow
Department of Biomedical Engineering
National University of Singapore
Singapore, Singapore

ISSN 1680-0737 ISSN 1433-9277 (electronic)
IFMBE Proceedings
ISBN 978-3-032-04371-9 ISBN 978-3-032-04372-6 (eBook)
https://doi.org/10.1007/978-3-032-04372-6

This Springer imprint is published by the registered company Springer Nature Switzerland AG
The registered company address is: Gewerbestrasse 11, 6330 Cham, Switzerland

Preface

The 18th International Conference on Biomedical Engineering (ICBME 2024), jointly convened by the Department of Biomedical Engineering of the National University of Singapore and the Biomedical Engineering Society (Singapore), welcomed more than 700 delegates from 34 countries to the NUS campus from 9 to 12 December 2024. With the continued endorsement of the International Federation for Medical and Biological Engineering and generous support from various sponsors, the four-day scientific programme showcased 194 distinguished, plenary, keynote and invited lectures as well as 128 special symposia and regular sessions. Collectively, these activities mapped the frontiers of biomedical engineering, from advanced biomaterials and intelligent sensing to computational physiology and neuroengineering, while nurturing the dialogue that has long characterised the ICBME series.

This Proceedings volume assembles ten peer-reviewed papers that collectively showcase the field's breadth, methodological rigour and translational ambition. The contributions range from digital-health innovations, exemplified by a multi-channel wearable auscultation vest that streams respiratory soundscapes for cloud-based diagnostics, to resource-optimised implant design such as cost-effective 3D-printed PMMA cranial plates tailored for low- and middle-income settings. They further include computational and experimental studies that illuminate the spatiotemporal release of therapeutic ions from biodegradable scaffolds, alongside complementary advances in biomechanics, neuroengineering, medical imaging and regenerative medicine. Viewed together, these works underscore how convergent expertise in materials science, data science and clinical insight is accelerating the translation of biomedical engineering research into tangible improvements in human health.

On behalf of the Organising and Scientific Committees, we express our deepest gratitude to the International Advisory Board, our international and local reviewers, speakers, exhibitors, sponsors and, most importantly, every participant whose curiosity and collegiality animated the discussions in Singapore. We are grateful to Nature Springer for their meticulous stewardship of this volume and to the many anonymous reviewers whose timely assessments ensured its quality. It is our sincere hope that these pages will inform, inspire and catalyse new collaborations, fortifying the community's shared mission to advance human health through biomedical engineering.

Chwee Teck Lim
Hwa Liang Leo
Raye Chen-Hua Yeow
Wei Seong Toh
Andrew Holle
Changsheng Wu

Organisation

Organising Committee

Chair

Chwee Teck Lim	National University of Singapore

Co-chair

Hwa Liang Leo	National University of Singapore

Advisor

James Goh	National University of Singapore

Secretary

Han Wei Hou	Nanyang Technological University

Treasurer

Peck Ha Tan	Ngee Ann Polytechnic

Scientific Chair

Raye Yeow	National University of Singapore

Scientific Co-chairs

Changsheng Wu	National University of Singapore
Andrew Holle	National University of Singapore
Wei Seong Toh	National University of Singapore

Sponsorship/Exhibition/Industrial Track

Gordon Xiong	Biomedical Engineering Society (Singapore)
Rex Tan	Aevice Health

Technical and Social Visits/Special Programmes

Jennifer Young	National University of Singapore
Peck Ha Tan	Ngee Ann Polytechnic

Clinical Track

Liang Zhong	Duke-NUS Medical School

Publicity

Alberto Corrias	National University of Singapore
Eliza Fong	National University of Singapore

Special Programmes

Eliza Fong	National University of Singapore
Andy Tay	National University of Singapore
Yuxin Liu	National University of Singapore
Yueming Jin	National University of Singapore
Aishwarya Bandla	Paxman Coolers Ltd.

International Advisory Board

Metin Akay	University of Houston
Zhenan Bao	Stanford University
Rashid Bashir	University of Illinois Urbana-Champaign
Freddy Boey	City University of Hong Kong
Shu Chien	University of California, San Diego
Tejal Desai	Brown University
David Elad	Tel Aviv University
Yubo Fan	Beihang University
Dietmar Hutmacher	Queensland University of Technology
Taeghwan Hyeon	Seoul National University
Roger Kamm	Massachusetts Institute of Technology
Ratko Magjarević	University of Zagreb
John A. Rogers	Northwestern University
Josep Samitier	Institute for Bioengineering of Catalonia

Metin Sitti	Koç University
Takao Someya	University of Tokyo
Fong-Chin Su	National Cheng Kung University
Rocky Tuan	Chinese University of Hong Kong
David Weitz	Harvard University
Tien Yin Wong	Tsinghua University
Lisa Xu	Shanghai Jiao Tong University
Yuanting Zhang	Chinese University of Hong Kong

International Scientific Committee

Taiji Adachi	Kyoto University
Noor Azuan Bin Abu Osman	University of Malaya
Anthony Bull	Imperial College London
Warren Chan	University of Toronto
Barbara Chan	Chinese University of Hong Kong
Wei Chen	University of Sydney
Jinwook Choi	Seoul National University
Jonathan Cooper	University of Glasgow
Tianghong Cui	University of Minnesota
Hyeong-Kim Dae	Seoul National University
Qionghai Dai	Tsinghua University
Dennis Discher	University of Pennsylvania
Yanan Du	Zhejiang University
Ross Ethier	Georgia Institute of Technology
Wei Gao	California Institute of Technology
Amit Gefen	Tel Aviv University
Guy Genin	Washington University in St. Louis
Keisuke Goda	University of Tokyo
Justin Gooding	University of New South Wales
Marie-Christine Ho Ba Tho	University of Technology of Compiègne
I-Ming Hsing	Hong Kong University of Science & Technology
Fatimah Ibrahim	University of Malaya
Ciprian Iliescu	National University of Singapore
Takuji Ishikawa	Tohoku University
Timo Jämsä	University of Oulu
Xingyu Jiang	Southern University of Science and Technology
Chulhong Kim	Pohang University of Science and Technology
Piotr Ladyzynski	Nalecz Institute of Biocybernetics & Biomedical Engineering
Andrew Laine	Columbia University
Peter Lee	University of Melbourne

Abraham Lee	University of California, Irvine
Vincent Lee	National University of Singapore
Song Li	University of California, Los Angeles
Zhi-Pei Liang	University of Illinois Urbana-Champaign
Feng-Huei Lin	National Taiwan University
Kang-Ping Lin	Chung Yuan Christian University
Yang Liu	Carnegie Mellon University
Peng Liu	Shanghai Jiao Tong University
Mian Long	Chinese Academy of Sciences
Takeo Matsumoto	Nagoya University
Paul Park	Massachusetts General Hospital
Leandro Pecchia	University of Warwick
Ichiro Sakuma	University of Tokyo
Amy Shen	Okinawa Institute of Science & Technology
Jennifer Shin	Korea Advanced Institute of Science & Technology
Anderson Shum	University of Hong Kong
Xiaolei Song	Southeast University
Jackrit Suthakorn	Mahidol University
Madoka Takai	University of Tokyo
Shoji Takeuchi	University of Tokyo
Nitish Thakor	National University of Singapore
Ashvin Thambyah	University of Auckland
Yi-Chin Toh	National University of Singapore
Shanbao Tong	Shanghai Jiao Tong University
Raymond Tong	Chinese University of Hong Kong
Jaw-Lin Wang	National Taiwan University
May Wang	Georgia Institute of Technology
Justin Cooper-White	University of Queensland
Ed X. Wu	University of Hong Kong
Younan Xia	Georgia Institute of Technology
Feng Xu	Xi'an Jiaotong University
Feng Xue	Peking University
Evelyn Yim	University of Waterloo
Yong Zhang	National University of Singapore
Grace Zhao	University of Connecticut
Xuanhe Zhao	Massachusetts Institute of Technology

Local Scientific Committee

Raye Yeow	National University of Singapore
Ali Asgar Bhagat	National University of Singapore

Toyama Yusuke	National University of Singapore
Yong Zhang	National University of Singapore
Wenting Zhao	National University of Singapore
Liang Zhong	National Heart Centre Singapore

Contents

Towards Prediction of Peri-Prosthetic Fractures: A Comprehensive Finite Element Model of Uncemented Total Hip Arthroplasty

Vineet Seemala[1](✉), Mark A. Williams[1], Richard King[2], and Arnab Palit[1]

[1] WMG, University of Warwick, Coventry CV4 7AL, UK
vineet.seemala@warwick.ac.uk
[2] Department of Trauma and Orthopaedics, University Hospitals Coventry and Warwickshire NHS Trust, Clifford Bridge Rd, Coventry CV2 2DX, UK

Abstract. Peri-prosthetic fractures (PPF) are a major concern in Total Hip Arthroplasty (THA), with intraoperative-PPF occurring more frequently than postoperative-PPF. Although finite element analysis (FEA) has been used to assess THA, it often inadequately models surgical steps like implantation. This study develops a novel FEA framework to simulate the surgical steps of uncemented THA, aiming to predict IOPPF risk. Bone geometry was segmented, a neck osteotomy was performed in-silico, and a mesh convergence study was conducted. The femur was modelled with a 10-node-tetrahedral mesh, using an isotropic-bilinear-heterogeneous material with a plastic-strain-based-failure-criterion and element deactivation to mimic fracture. Boundary conditions included fixation of the distal femur and implant path with local-adaptive remeshing. The FEA model was subsequently verified through a cadaver experiment. The FEA model successfully simulated the surgical process, with no high-strain regions observed during implantation with the surgeon-selected implant size. However, when a larger implant size was used in the simulation, high-strain regions indicative of IOPPF risk were identified, confirming the model's ability to predict fracture risk based on implant size. The developed FEA framework can assist surgeons in pre-surgical planning, particularly in implant selection, positioning, and predicting PPF risk. Future improvements will aim to refine and validate the model.

Keywords: Total Hip Replacement (THR) · Peri-Prosthetic Fracture (PPF) · Finite Element Analysis (FEA)

1 Introduction

Total Hip Arthroplasty (THA) is an effective surgical intervention for alleviating pain and restoring mobility in patients with hip osteoarthritis. Hip implants are categorised into cemented and uncemented types based on their bonding mechanisms. Although modern cemented implants perform well, limitations such as poor tensile strength and the risk of osteolysis have led to increased use of uncemented implants [1, 2]. In the UK, the use of cemented hips nearly halved from 2006 to 2021, while uncemented implants

C. T. Lim et al. (Eds.): ICBME 2024, IFMBE Proceedings 132, pp. 1–11, 2025.
https://doi.org/10.1007/978-3-032-04372-6_1

increased by almost 2.5 times during the same period [3]. By 2030, the number of THA procedures in young adults is expected to increase fivefold [4], with over 80% of these patients receiving uncemented implants [3].

Approximately 10% of these procedures require revision surgery within 15 years due to intraoperative and postoperative complications [5, 6]. Periprosthetic fractures (PPF) significantly impact surgical outcomes, increasing aseptic loosening risk by 7.2 times, contributing to post-revision complications, and elevating the likelihood of postoperative PPF [7]. The risk of intraoperative PPF (IOPPF) is over three times higher for uncemented implants compared to cemented ones, often occurring during broaching and uncemented implantation [8–11]. Predicting PPF risk remains challenging due to variations in patient bone geometry and morphology [12–15].

Finite Element Analysis (FEA) has been instrumental in studying bone-implant interactions and assessing uncemented THA performance [16–21]. While FEA has been used to evaluate postoperative implant stability [16, 20], it falls short in assessing intraoperative PPF (IOPPF) risk, despite IOPPF occurring more frequently than postoperative PPF (POPPF). Existing FEA models simplify the uncemented THA process, representing press-fitting as either volumetric expansion [17] or opposing forces [18, 19], overlooking the intricate interactions between the bone, broach, and implant. To the authors' knowledge, no study has modelled IOPPF risk during uncemented THA using FEA.

Therefore, the aim of this study is to develop a FEA framework that mimics the surgical steps of uncemented THA to help guide pre-surgical planning. To achieve this aim, the following objectives have been set:

A. Define the implant path into the femur to establish the boundary conditions of the FEA simulation.
B. Develop a FEA framework for the uncemented THA surgical process, modelling the surgical steps.
C. As a case study, perform uncemented THA on a cadaveric femur sample and compare the developed FEA model with the experiment.

2 Methods and Materials

The following sections detail the steps involved in setting up a FEA of the uncemented THA surgical process, followed by a case study to compare the developed FEA with the experiment.

2.1 Geometry Acquisition

The patient-specific femur geometry was obtained through a medical CT scan. Similarly, the geometries of the implant were acquired using a laser scanner (Nikon ModelMaker H120, mounted on a Nikon MCAx S portable CMM arm) and a μCT scan (Tescan Unitom XL).

2.2 Segmentation and Alignment

The femurs and femoral implants from CT scans were segmented using watershed-based techniques [22] in Avizo 3D 2021.2 (Thermo Fisher Scientific, Germany). Surface mesh generation followed, with edge length optimised through a mesh convergence study. A coordinate system for the femur was established per ISB recommendations [23], using anatomical landmarks (femoral head centre, lateral epicondyle, medial epicondyle) from the CT scans [24] and aligned with the global coordinate system. Finally, the neck osteotomy was performed in silico at the piriformis muscle insertion, following surgical guidelines.

2.3 Implant Path Definition

The centroid axis of the femoral canal was defined using its surface in Matlab 2022b (The MathWorks, Inc., Natick, MA), as shown in Fig. 1. The tangential direction of this axis was then determined. Following surgical planning guidelines, the implant axis was established, with the path defined by aligning it parallel to the tangent of the femoral canal centroid axis at the insertion depth, considering prior tangential directions. The implant direction at the jth increment for n increments of implant insertion was then determined as follows

$$\overline{a_j} = \frac{\left(\frac{\sum_{i=1}^{j} \overline{a_i}}{j}\right)}{\left|\frac{\sum_{i=1}^{j} \overline{a_i}}{j}\right|} j \in [1, n] \tag{1}$$

where $\overline{a}_j$ is implant axis direction, n is the number of increments, and j is the current increment.

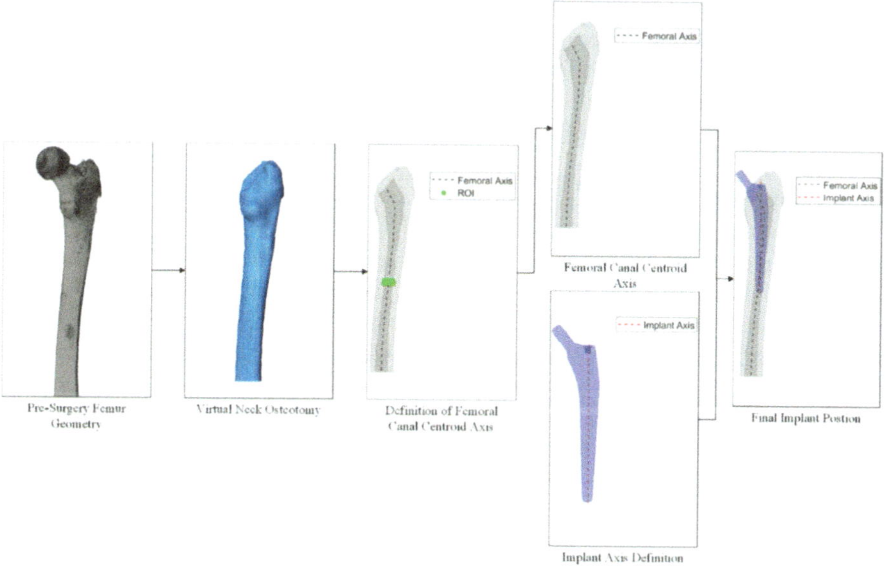

Fig. 1. Implant Path Definition Framework

2.4 Subject Specific FEA Model

A geometry-based FE model was developed using Marc Mentat 2023.3 (MSC Software Corporation, Hexagon, Sweden), with geometries derived from CT scans, as illustrated in Fig. 2. The bone geometry was meshed using 10-node tetrahedral elements, ensuring the maximum edge length stayed below the threshold determined by the mesh convergence study. The FEA was modelled under quasi-static loading conditions, with a simulation timeframe that allowed mass inertia to be neglected.

Material Property Definition
Bone density was measured using density calibration, where CT scan intensities were converted to density values following Seemala et al. [21, 25]. Material properties at each voxel were then evaluated using appropriate empirical correlations. Young's modulus [26] and yield stress [27] were determined based on the following correlations (Eq. (2)), with Poisson's ratio set to 0.3 and a maximum plastic strain failure criterion of 0.01 [28]. The heterogenous bone material properties were assigned to each integration point using the '*hooklw*', '*wkslp*', and '*ufail*' subroutines. To simulate bone failure and fracture, the element deactivation option in Marc Mentat was utilised.

$$E = 6580\rho_{ash}^{1.49}$$

$$\sigma_{ys} = \begin{cases} 137(\rho_{ash})^{1.88} for \rho_{ash} < 0.317g/cc \\ 114(\rho_{ash})^{1.72} for \rho_{ash} \geq 0.317g/cc \end{cases} \quad (2)$$

Boundary Condition Definition
The boundary conditions were defined as follows: the distal femur was fixed to prevent displacement and rotation, replicating the patient's leg being held during the surgery. Additionally, the implant was displaced and rotated into the femur along the path outlined in Sect. 2.3, simulating its progression under mallet strikes during implantation.

Mesh Convergence Study
An ROI near the lesser trochanter, characterised by high bone-implant contact [29], was selected. A mesh convergence study was conducted to determine the optimal maximum edge length for the 10-node tetrahedral mesh of the bone geometry, ensuring displacement and stress convergence. Seven FEA models with maximum edge lengths of 8 mm, 6 mm, 4 mm, 3 mm, 2 mm, 1.25 mm, and 1 mm were compared. The results guided the subsequent meshing for FEA, adhering to the determined maximum edge length. A local adaptive remeshing was applied to the elements in the four-cylinder region surrounding the broach and implant, with two levels of refinement, as shown in Fig. 2.

2.5 Case Study

The FEA model developed in this study (Fig. 2) was verified through uncemented THA on cadaveric femur sample, with the experimental findings compared to the FEA simulation results. A single cadaveric femur sample, from a 70-year-old male (right femur), was

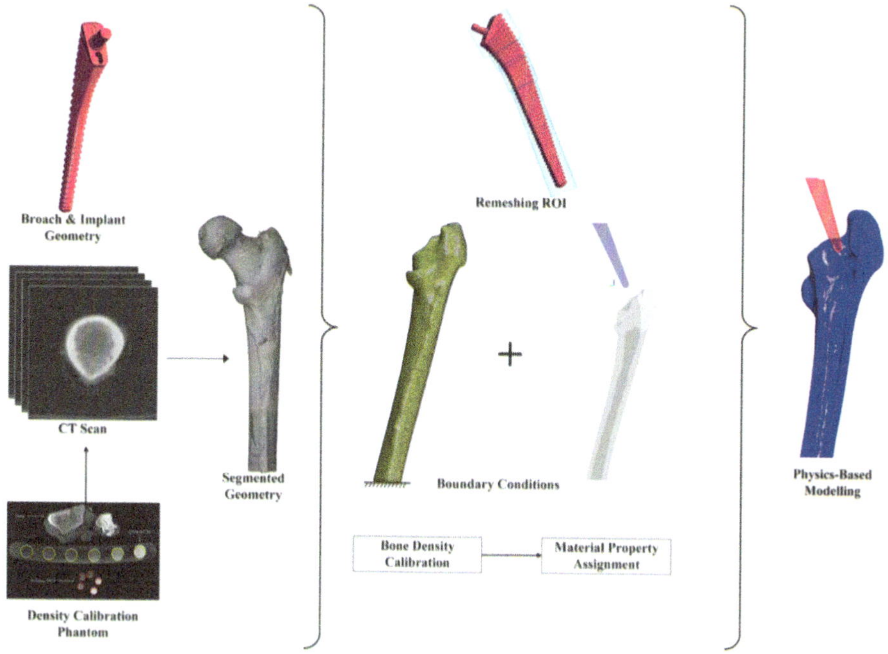

Fig. 2. FEA Workflow Framework

used in this study. The femur was stored at −20 °C and thawed to room temperature 24 h before the experiment. Ethical approval was obtained from the Biomedical & Scientific Research Ethics Committee (BSREC) at the University of Warwick (Ref: BSREC 66/22-23) and the Research and Development department at University Hospitals Coventry and Warwickshire (UHCW) NHS Trust (Ref: GF0503). An orthopaedic surgeon with over 25 years of experience performed the preparation and THA procedures. A standard Metafix (Corin, Cirencester, Gloucestershire, UK) implant 4, selected by the surgeon, was used for the femur sample. Detailed experimental setup information is available in the authors' previous work [21, 29]. Figure 3 the experimental setup for performing uncemented THA on the cadaveric femur sample.

3 Results

3.1 Mesh Convergence Results

The results of the mesh convergence study are shown in Fig. 4. The stress convergence is depicted in Fig. 4a, while Fig. 4b illustrates the displacement convergence based on different edge lengths, as described in Sect. 2.4. The expected convergence curves were observed. The ROI used for the mesh convergence study is presented in Fig. 4c, and the FEA is displayed in Fig. 4d. A maximum edge length of 2 mm was selected based on the mesh displacement and stress convergence, ensuring good computational efficiency while maintaining displacement and stress convergence with a 10-node tetrahedral element.

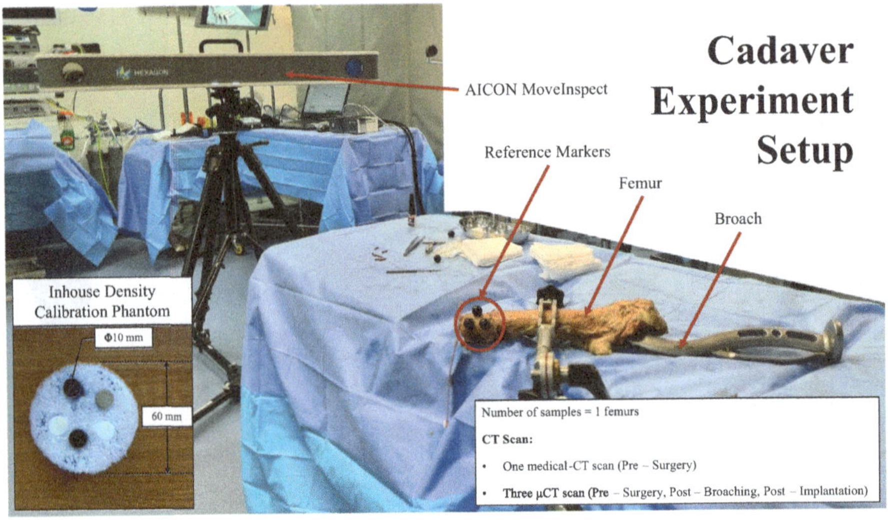

Fig. 3. Cadaver Experiment Setup

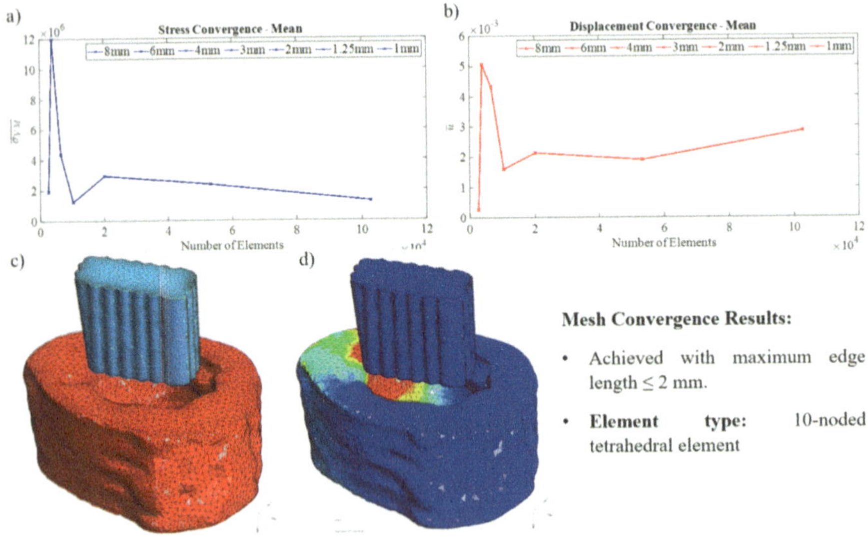

Fig. 4. Mesh Convergence Results: (a) Stress convergence, (b) Displacement convergence, (c) Mesh convergence in the region of interest (ROI), (d) Finite Element Analysis (FEA) results.

3.2 Successful Implantation FEA

Figure 5 presents the FEA results obtained using the FEA workflow developed in this study, considering the implant size selected by the orthopaedic surgeon during the cadaver experiment. The equivalent strain (ε_{eq}) is depicted in the colour map. The strain induced after implantation is shown in both the anterior and posterior views, along with the final implant position (Fig. 5). No high-strain regions were observed, indicating no risk of periprosthetic fracture during uncemented implantation for the selected implant size.

3.3 Peri-Prosthetic Fracture Prediction from FEA

Figure 6 shows the PPF prediction based on the FEA results obtained using the FEA workflow developed in this study, considering an implant size that is two sizes larger than the one selected during the cadaver experiment. The equivalent strain (ε_{eq}) is depicted in the colour map. The strain induced after implantation is illustrated in both the anterior and posterior views (Fig. 6). Large high-strain regions were observed, indicating a risk of periprosthetic fracture during uncemented implantation, with the areas where PPF occurred highlighted in red for the larger implant size.

4 Discussion

The prediction of PPF risk is complicated by variations in patient bone geometry and morphology [12–15]. FEA has been utilised to study bone-implant interactions and postoperative implant stability [16, 20] but remains inadequate for assessing IOPPF risk in the current literature. Therefore, this study aimed to develop an FEA framework that replicates the surgical steps of uncemented THA to assist in pre-surgical planning. Key contributions include: A) Defining the implant path within the femur. B) Developing an FEA framework for uncemented THA to predict PPF. C) Comparing the FEA model with experimental results using cadaveric femurs.

To the best of the authors' knowledge, no prior study has modelled the risk of IOPPF during uncemented THA using FEA. Existing FEA models simplify the uncemented THA process by representing press-fitting as either volumetric expansion [17] or opposing forces [18, 19], thereby neglecting the complex interactions between the bone, broach, and implant. Consequently, in this study, the implant path into the bone—mimicking its progression under mallet strikes during implantation—was defined as outlined in Sect. 2.4. Following surgical planning guidelines, the implant axis was established, with the path determined by aligning it parallel to the tangent of the femoral canal centroid axis at the insertion depth, while also considering prior tangential directions. This defined implant path serves as the boundary condition for the FEA, enabling more accurate interaction between the implant and femur during the surgical procedure and enhancing the prediction of IOPPF risk during uncemented THA.

A physics-based modelling workflow was developed for the first time to simulate uncemented THA (Fig. 2), replicating the surgical steps to the best of the authors' knowledge. The patient-specific inputs required for the modelling, as well as the boundary conditions that mimic the surgical conditions, were defined. The FEA model of the case study demonstrated that the workflow, developed in this study, reflected the implant size selected by the orthopaedic surgeon during the cadaver experiment. No high-strain regions were observed, indicating no risk of periprosthetic fracture during uncemented implantation for the selected implant size, as shown in Fig. 5. However, when an implant two sizes larger than the one used in the cadaver study was modelled, the FEA predicted

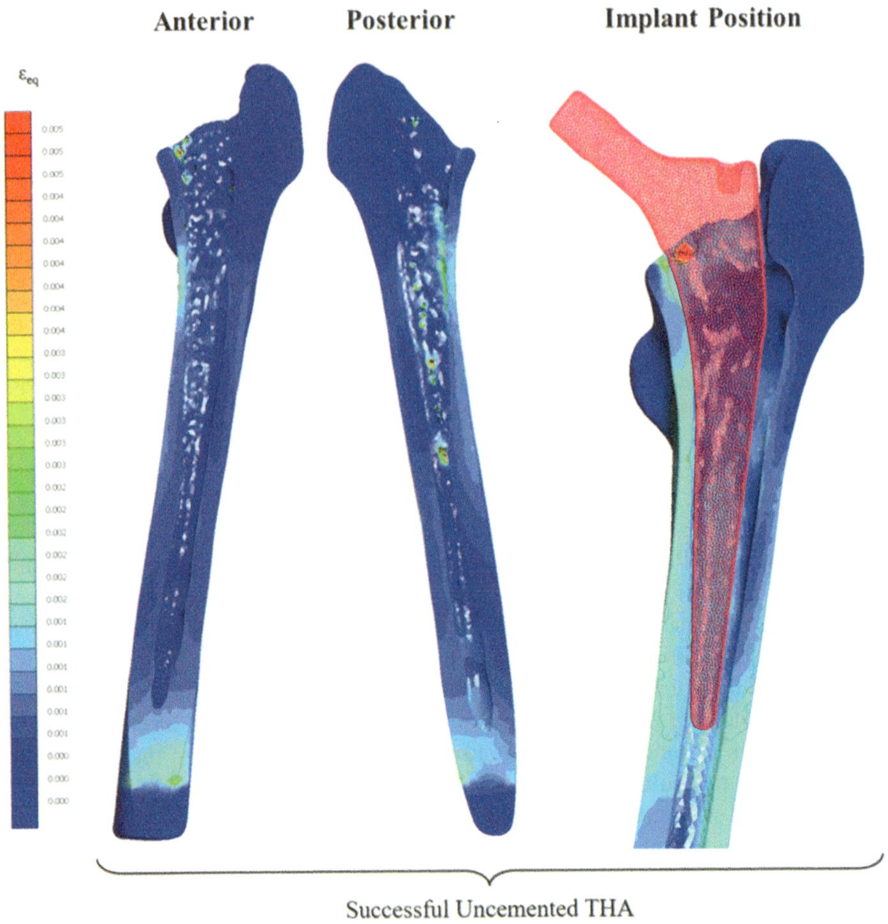

Fig. 5. Uncemented THA FEA: Implant size used in the cadaver experiment as determined by the orthopaedic surgeon, with no incidence of peri-prosthetic fracture. The colourmap represents the equivalent strain.

the occurrence of IOPPF. Large high-strain regions were observed, indicating a risk of periprosthetic fracture during uncemented implantation, as shown in Fig. 6. Therefore, this physics-based model of uncemented THA can assist orthopaedic surgeons in making more informed decisions during pre-surgical planning, particularly in selecting the implant size and position to ensure no risk of PPF and enhance implant stability.

However, a few limitations and improvements are required for this study. Firstly, the complete uncemented THA surgical steps, including broaching, should be modelled to predict a comprehensive IOPPF risk, which would improve the IOPFF prediction and aid in guiding pre-surgical planning. Secondly, the PPF prediction from the developed physics-based modelling should be validated through an experimental study, including the prediction of PPF load and position. Thirdly, the developed physics-based model should be compared with the current state-of-the-art to assess the differences. Therefore,

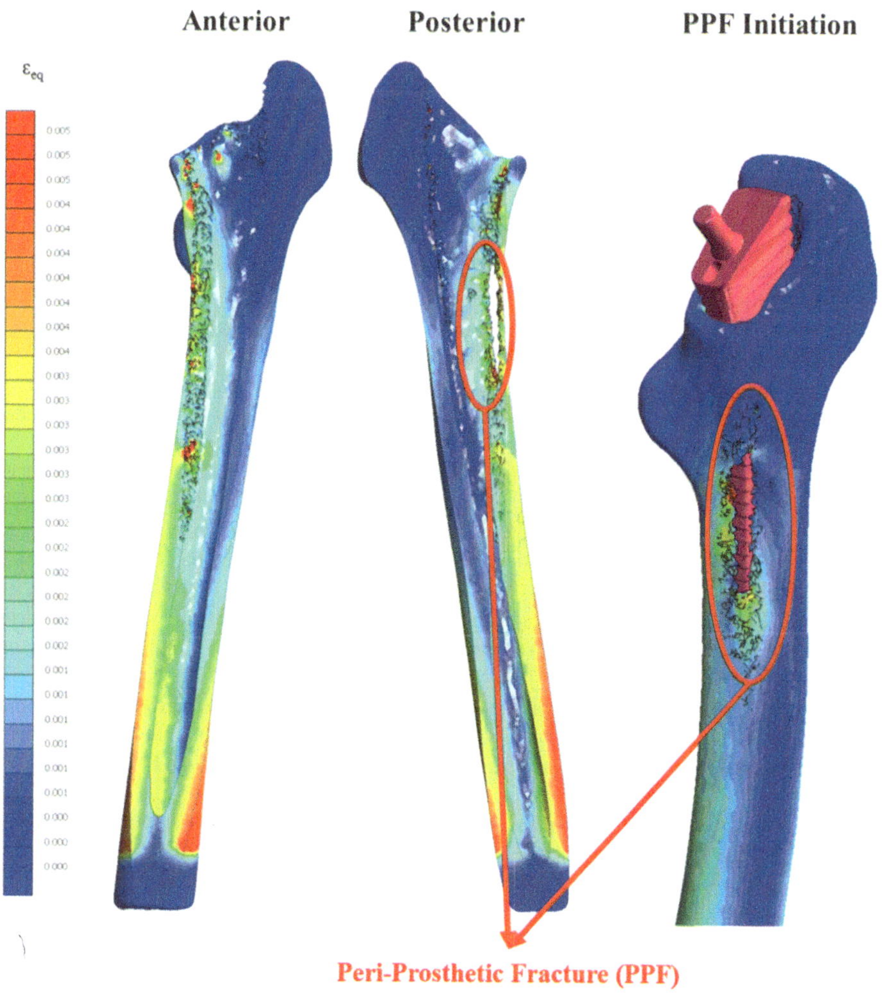

Fig. 6. Peri-Prosthetic Fracture Prediction from FEA: Implant size two sizes larger than that used in the cadaver experiment, demonstrating the incidence of PPF, with PPF regions highlighted in red. The colourmap represents the equivalent strain (color figure online).

these limitations should be addressed in future studies to improve uncemented THA modelling.

5 Conclusion

This study presents a novel FEA framework for simulating the surgical steps of uncemented THA, particularly focusing on the risk of IOPPF. By modelling the interaction between the bone and implant during surgery, the developed framework provides valuable insights into the potential PPF risk associated with implant selection and positioning. The case study demonstrated that the model successfully predicted the risk of IOPPF when an oversized implant was used. This approach can assist orthopaedic surgeons in making informed decisions during pre-surgical planning, although future studies are needed to refine the model further and validate its predictions through experimental comparisons.

References

1. Huddleston, H.D.: Femoral lysis after cemented hip arthroplasty. J. Arthroplasty **3**(4), 285–297 (1988)
2. Chareancholvanich, K., et al.: In vitro stability of cemented and cementless femoral stems with compaction. Clin. Orthop. Relat. Res. **394**, 290–302 (2002)
3. Ben-Shlomo, Y., et al.: National Joint Registry Annual Reports. The National Joint Registry 19th Annual Report 2022. National Joint Registry © National Joint Registry 2022, London (2022)
4. Kurtz, S.M., et al.: Future young patient demand for primary and revision joint replacement: national projections from 2010 to 2030. Clin. Orthop. Relat. Res. **467**(10), 2606–2612 (2009)
5. Deere, K., et al.: How long do revised and multiply revised hip replacements last? A retrospective observational study of the National Joint Registry. Lancet Rheumatol. **4**(7), e468–e479 (2022)
6. Evans, J.T., et al.: How long does a hip replacement last? A systematic review and meta-analysis of case series and national registry reports with more than 15 years of follow-up. Lancet **393**(10172), 647–654 (2019)
7. Lamb, J.N., et al.: Patient and implant survival following intraoperative periprosthetic femoral fractures during primary total hip arthroplasty: an analysis from the national joint registry for England, Wales, Northern Ireland and the Isle of Man. Bone Joint J. **101-B**(10), 1199–1208 (2019)
8. Lindberg-Larsen, M., et al.: Increased risk of intraoperative and early postoperative periprosthetic femoral fracture with uncemented stems. Acta Orthop. **88**(4), 390–394 (2017)
9. Bruggemann, H., et al.: Incidence and risk factors of intraoperative periprosthetic femoral fractures during primary total hip arthroplasty: 218,423 cases reported to the Norwegian Arthroplasty Register between 1987 and 2020. Acta Orthop. **93**, 405–412 (2022)
10. Holley, K., et al.: Periprosthetic fractures of the femur after hip arthroplasty: an analysis of 99 patients. HSS J. **3**(2), 190–197 (2007)
11. Hu, K., et al.: Periprosthetic fractures may be more likely in cementless femoral stems with sharp edges. Ir. J. Med. Sci. **179**(3), 417–421 (2010)
12. Carli, A.V., Negus, J.J., Haddad, F.S.: Periprosthetic femoral fractures and trying to avoid them: what is the contribution of femoral component design to the increased risk of periprosthetic femoral fracture? Bone Joint J. **99-b**(1 Supple A), 50–59 (2017)
13. Konow, T., et al.: Factors influencing periprosthetic femoral fracture risk. Bone Joint J. **103-B**(4), 650–658 (2021)
14. Ramavath, A., et al.: Postoperative periprosthetic femoral fracture around total hip replacements: current concepts and clinical outcomes. EFORT Open Rev. **5**(9), 558–567 (2020)
15. Abdul-Kadir, M.R., et al.: Finite element modelling of primary hip stem stability: the effect of interference fit. J. Biomech. **41**(3), 587–594 (2008)
16. Hennicke, N.S., et al.: Subject specific finite element modelling of periprosthetic femoral fractures in different load cases. J. Mech. Behav. Biomed. Mater. **126**, 105059 (2022)
17. Esposito, L., et al.: Towards a patient-specific estimation of intra-operative femoral fracture risk. Comput. Methods Biomech. Biomed. Eng. **21**(12), 663–672 (2018)

18. Petursson, T., et al.: Bone mineral density and fracture risk assessment to optimize prosthesis selection in total hip replacement. Comput. Math. Methods Med. **2015**, 162481 (2015)
19. Gargiulo, P., et al.: Assessment of total hip arthroplasty by means of computed tomography 3D models and fracture risk evaluation. Artif. Organs **37**(6), 567–573 (2013)
20. Hennicke, N.S., Kluess, D., Sander, M.: Influence of stem design parameters on periprosthetic femoral fractures examined by subject specific finite element analyses. Med. Eng. Phys. **119**, 104032 (2023)
21. Seemala, V., et al.: Medical vs MicroCT based finite element analysis : exploring the influence of bone heterogeneity and bone geometry. In: 2024 46th Annual International Conference of the IEEE Engineering in Medicine and Biology Society (EMBC), pp. 1–5 (2024)
22. Beucher, S., Meyer, F.: The morphological approach to segmentation: the watershed transformation. In: Mathematical Morphology in Image Processing, pp. 433–481. CRC Press (2018)
23. Wu, G., et al.: ISB recommendation on definitions of joint coordinate system of various joints for the reporting of human joint motion—part I: ankle, hip, and spine. J. Biomech. **35**(4), 543–548 (2002)
24. Palit, A., King, R., Pierrepont, J., Williams, M.A.: Development of bony range of motion (B-ROM) boundary for total hip replacement planning. Comput. Methods Programs Biomed. **222**, 106937 (2022)
25. Seemala, V., et al.: A micro-CT based cadaveric study investigating bone density changes during hip arthroplasty surgery. J. Orthop. Res. (2024)
26. Morgan, E.F., Bayraktar, H.H., Keaveny, T.M.: Trabecular bone modulus–density relationships depend on anatomic site. J. Biomech. **36**(7), 897–904 (2003)
27. Revel, M., et al.: A credible homogenized finite element model to predict radius fracture in the case of a forward fall. J. Mech. Behav. Biomed. Mater. **131**, 105206 (2022)
28. Miles, B., et al.: Subject specific finite element modeling of periprosthetic femoral fracture using element deactivation to simulate bone failure. Med. Eng. Phys. **37**(6), 567–573 (2015)
29. Seemala, V., et al.: Quantifying bone compaction and implant-bone contact in uncemented total hip arthroplasty through muCT and digital volume correlation: a cadaveric study. Comput. Biol. Med. **184**, 109474 (2024)

Computational Modeling of Right Coronary Artery Flow Impairment in Pulmonary Hypertension

Chenghan Cai[1], Jenny S. Choy[2], Ghassan S. Kassab[2], and Lei Fan[1](✉)

[1] Joint Department of Biomedical Engineering, Marquette University and Medical College of Wisconsin, Milwaukee, WI, USA
lei.fan@marquette.edu
[2] California Medical Innovations Institute, San Diego, CA, USA

Abstract. Pulmonary hypertension (PH) characterized by elevated mean pulmonary arterial pressure (mPAP) impairs coronary flow that leads to right ventricular (RV) failure, however, the underlying mechanism that controls the transition from compensated to decompensated RV failure remains poorly understood. To address this, we developed a computational model that integrates coronary perfusion in the major coronary arteries with a biventricular mechanics model in a closed-loop systemic and pulmonary circulations, to quantify the effects of acute and chronic PH on heart function and coronary blood regionally and transmurally cross the myocardium. The model predicts a reduction in coronary flow in the right ventricular free wall (RVFW) and septum during acute PH due to elevated intramyocardial pressure in the RV, which is higher in the endocardium than that in the epicardium. The model also predicted that the coronary flow in the left ventricular free wall (LVFW), septum, and RVFW is reduced in chronic PH due to diminished perfusion pressure. The computational model predictions suggest that the impairment of right coronary flow in chronic PH increases ischemic risk through reduced left coronary perfusion, which further affects the left ventricular (LV) function. These findings indicate the effects of PH on regional coronary flow may determine the transition from compensated to decompensated RV failure.

Keywords: Computational modeling · pulmonary hypertension · right coronary perfusion

1 Introduction

Right heart failure (RHF) is a leading cause of death in patients with pulmonary hypertension (PH), which is associated with a resting mean pulmonary arterial pressure (mPAP) greater than 20 mmHg (Condon et al., 2019). The prognosis of patients with PH remains poor, with approximately 10–15% mortality within 1 year of the diagnosis despite the administration of advanced therapy (McLaughlin et al., 2009). With the progression of PH, a sustained increase in right ventricular (RV) afterload results in an increased RV

C. T. Lim et al. (Eds.): ICBME 2024, IFMBE Proceedings 132, pp. 12–24, 2025.
https://doi.org/10.1007/978-3-032-04372-6_2

workload and potentially hypertrophy. Although increased RV workload might necessitate enhanced RV myocardial perfusion and oxygen consumption, a reduction in systolic flow in the right coronary artery (RCA) has been reported in PH (Haddad et al., 2008; Manohar et al., 1981). RV ischemia is thought to be due to RV remodeling. However, it remains challenging to elucidate the effects of each individual mechanism in PH on coronary perfusion. As such, our understanding of PH and how it affects RV mechanics and coronary perfusion remains incomplete.

The development of computational models will provide an insight to determine the interplay between biventricular mechanics and coronary perfusion under pathophysiological conditions (Fan, 2019; Richardson et al., 2021). Currently, the majority of computational models are focused on the left ventricle (LV) (Nordsletten et al., 2011; Shavik et al., 2021) whereas the right side of the heart has received relatively little attention. In particular, some studies developed thick-walled spherical segments for ventricular mechanics (Lumens et al., 2009), which have been applied to study the effects of acute and chronic PH on ventricular mechanics. Other studies developed finite element (FE) models that account for the 3D geometry of the biventricular unit to allow a more detailed representation of regional RV mechanics (Augustin et al., 2021; Kakaletsis et al., 2023; Mendiola et al., 2023; Shavik et al., 2019). These models can provide detailed insights into the local distribution of stress and strain in the left ventricular free wall (LVFW), right ventricular free wall (RVFW), and septum. Right coronary perfusion and myocardium-vessel interaction in the RV, however, have been neglected (Fan et al., 2024b).

To address these limitations, we developed a computational framework consisting of a biventricular FE model coupled with coronary perfusion in a closed-loop lumped parameter representation of the systemic and pulmonary circulations. The mathematical model was calibrated based on clinical measurements of the LV and RV pressure volume (PV) loops from a healthy human. The model is applied to simulate acute and chronic PH and quantify their effects on coronary perfusion as well as biventricular mechanics.

2 Methods

The computational framework consists of a FE model of the biventricular unit that is coupled in a closed-loop system to **1)** models of coronary microvasculature networks in the perfusion territories of the RCA, left anterior descending (LAD) artery and left circumflex artery (LCx), **2)** a lumped model of the systemic and pulmonary circulations (Fig. 1a). There are a total of 9 coronary microvasculature networks located at the endocardial, mid-wall, and epicardial regions of the RVFW, interventricular septum, and LVFW. The RCA, LAD and LCx perfusion territories are associated with the 3 networks (at endocardium, mid-wall and epicardium) located at the RVFW, interventricular septum and LVFW, respectively.

2.1 Cardio-Pulmonary Circulation

The systemic and pulmonary circulations each consist of four cardiovascular components that are associated with the ventricles, atria, arteries, and veins, respectively. Conservation of blood volume in these circulations requires the net change of inflow and outflow rates of each compartment to be equal to its rate of change of the volume. Blood flow rates in the arteries and veins are determined by each segment's resistance and the pressure difference between the two connecting storage compartments. Pressures in the storage compartments of the peripheral (proximal and distal) arterial and venous compartments in the systemic and pulmonary circulations are represented by the prescribed resting volume and compliance of each compartment, respectively. The pumping characteristics of the right atrium (RA) and left atrium (LA) are represented by a time-varying elastance model (Fan et al., 2022, 2021b, 2020).

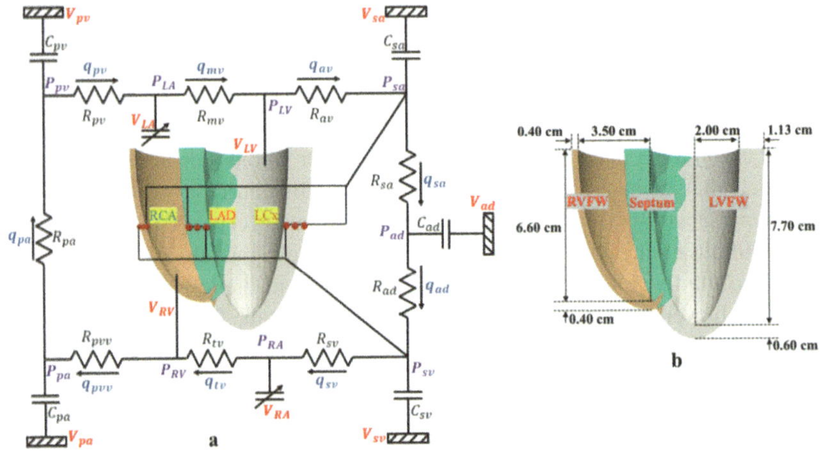

Fig. 1. a. Schematic of the closed-loop computational framework that couples the biventricular FE model with coronary microvasculature networks at endocardial, mid-wall, and epicardial regions of perfusion territories associated with the RCA, LAD and LCx (each red dot represents a coronary network); **b.** Dimension of the idealized biventricular geometry (color figure online).

2.2 Finite Element Biventricular Mechanics

The relationship between LV and RV pressures and volumes was described using a FE model of the biventricular unit (Fan et al., 2024a, 2023). Pressure-volume relationships of the LV and RV were determined from minimization of the Lagrangian function by taking the first variation of the function.

The geometry of the biventricular unit was idealized as the fusion of 2 truncated ellipsoids representing the LV and RV. Dimensions of the biventricular unit were prescribed based on those found in a healthy human heart (Ho and Nihoyannopoulos, 2006; Ryan and Archer, 2014) (Fig. 1b). Helix angle associated with the myofiber direction $\boldsymbol{e}_{f0}$ was prescribed to vary transmurally from 60° at the endocardium to −60° at the epicardium across the biventricular wall based on experimental measurements (Streeter et al., 1969).

The stress tensor $\boldsymbol{P}$ in each ventricle is decomposed additively into a passive component $\boldsymbol{P}_p$ and an active component $\boldsymbol{P}_a$ (i.e., $\boldsymbol{P} = \boldsymbol{P}_p + \boldsymbol{P}_a$). An active stress formulation

$\boldsymbol{P}_a$ was used to describe the mechanical behavior of the myocardial tissue in the cardiac cycle. The passive stress tensor was defined by $\boldsymbol{P}_p = dW/d\boldsymbol{F}$, where $\boldsymbol{F}$ is the deformation gradient tensor and W is a strain energy function of a Fung-type transversely-isotropic hyperelastic material (Guccione et al., 1991).

2.3 Coronary Blood Flow

Each coronary microvasculature network consists of 400 vessels that are pruned from a previously reconstructed coronary network as described in (Fan et al., 2020; Namani et al., 2018) and parameters used in coronary flow network model is based on (Namani et al., 2018). Briefly, flow in each vessel is represented by a three-element Windkessel model consisting of two identical nonlinear resistors and one nonlinear capacitor, which are functions of the vessel radius. The vessel radius depends on the transvascular pressure of each vessel, which is defined to be the difference between the vessel pressure and the intramyocardial pressure (IMP) that varies regionally. Mathematical formulation of the coronary microvasculature network (based on mass conservation) forms a nonlinear system of ordinary differential equations (ODEs) in terms of the unknown vessel pressure in the middle of each vessel. The ODEs were solved using the backward differentiation formula in CVODE (http://www.llnl.gov/casc/sundials/).

2.4 Myocardium-Vessel Interaction

FE model of the biventricular unit and the coronary microvasculature networks are coupled bidirectionally at different levels in the framework. ***First***, the proximal arterial and venous pressures in the circulatory system serve as pressure boundary conditions in each coronary network. Total inlet and outlet coronary flow from the RCA, LAD, and LCx networks computed with these pressure conditions are fed back into the lumped parameter model of the systemic circulation at each time step in a closed-loop. ***Second***, IMP in the coronary microvasculature networks is prescribed using the local value of Lagrange multiplier associated with the incompressibility constraint from the FE model as in previous studies (Fan et al., 2021c, 2021a). We note that the IMP is different for each network located at different transmural locations and different regions of the biventricular unit.

2.5 Simulation Cases

The computational framework was first calibrated to produce physiological cavity pressure and volume waveforms of the LV and RV, as well as PAP resembling those found in healthy humans (***baseline***). The calibrated model was then applied to simulate PH without remodeling (***acute PH***) by increasing the pulmonary arterial resistance, R_{pa}, and reducing the preload, V_{sv0} (Hoeper and Granton, 2011) to keep the ratio of RV end-diastolic volume (EDV)/LV EDV below 1.5 (Finsberg et al., 2019). Thereafter, we simulate PH with remodeling (***chronic PH***) by increasing the RVFW thickness from 3 mm to 9 mm (Hilde et al., 2013). In all simulations, approximately 14,000 (13,694 for baseline) elements were used for discretizing the geometry. Simulations were performed using Medical College of Wisconsin High Performance Computing Cluster with

12 processors with each having a base clock speed of 2.5 GHz. The time step was set as 1 ms in the cardiac cycle and the simulation was terminated when the results reached a steady periodic state.

3 Results

3.1 Baseline

For the baseline case simulating hemodynamics and ventricular function of those found in healthy human subjects, the model predicted mPAP is 14 mmHg, peak RA pressure 3 mmHg, and the peak LV and RV pressures 125 and 28 mmHg, respectively (Fig. 2a). Ejection fraction (EF) is 54% for both cavities and the LV and RV end-diastolic volume (EDV) 99 and 100 mL, respectively, which are in the physiological range for normal humans (Bernard et al., 2017) (Fig. 2b–c). The transmural variation of peak IMP across the ventricular wall was nearly linear (Fig. 2d). Specifically, peak IMP reduces from 128 to 24 mmHg and 39 to 18 mmHg from endocardium to epicardium in the LVFW and RVFW, respectively. In the septum, peak IMP reduces from 133 mmHg in the LV endocardium to 36 mmHg in the RV endocardium. The gradient of IMP variation is smaller in the RVFW than that in the LVFW and septum. Transmural variation of the peak IMP to cavity pressure ratio reduces from endocardium to epicardium in the LVFW and RVFW. Peak ratio was higher in the RVFW endocardium (1.38) than that in the LVFW endocardium (1.03) (Fig. 2e).

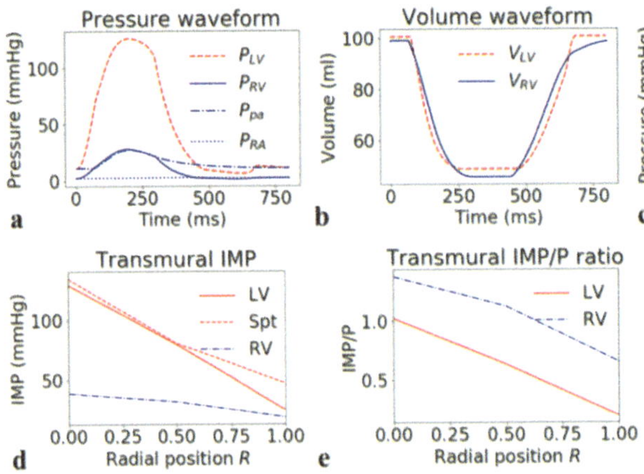

Fig. 2. Model predicted **a.** Pressure waveforms; **b.** Volume waveforms; **c.** PV loops; **d.** transmural distribution of IMPs; **e.** transmural distribution of IMP and its cavity pressure ratio. For LVFW and RVFW, R = 0 and 1 correspond to endocardium and epicardium, respectively. For septum, R = 0 and 1 correspond to LV endocardium and RV endocardium, respectively.

Because of the transmural variation of IMP, total coronary flow varies transmurally. Flow rate waveforms at different transmural locations of the RVFW, septum, and LVFW show that flow rates at endocardium of each region are the lowest due to higher IMP at endocardium (Fig. 3a–c). Since the IMP in the RVFW is significantly lower than perfusion pressure, coronary flow rate at different transmural locations of the RVFW increases during systole (Fig. 3a). Total coronary flow over a cardiac cycle increases transmurally from 0.26 mL at the endocardium to 0.46 mL at the epicardium of the LVFW and from 0.25 mL at the endocardium to 0.27 mL at the epicardium of the RVFW. At the septum, total coronary flow increases from 0.53 mL at the LV endocardium to 0.89 mL at the RV endocardium. The model predicted that total coronary flow over a cardiac cycle in each microvascular network agrees with the measurements from healthy humans (Sakamoto et al., 2013). Furthermore, the gradient of the transmural variation of total coronary flow in the RVFW is lower than that of the LVFW due to the lower gradient in the transmural distribution of IMP (Fig. 3d).

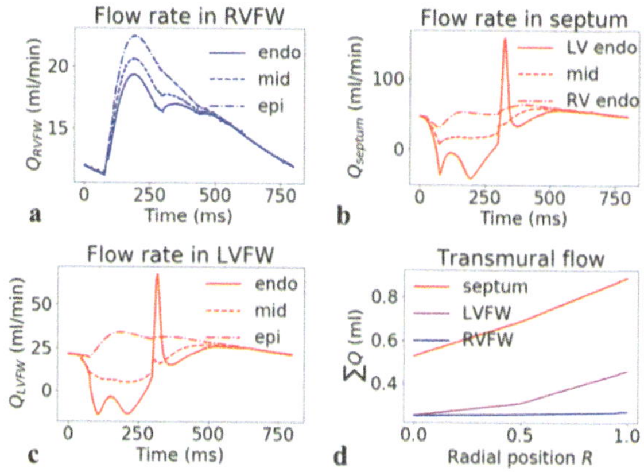

Fig. 3. Model predicted flow rate waveforms at transmural locations in the **a.** RVFW; **b.** septum; **c.** LVFW. **d.** Transmural distribution of total coronary flow. For LVFW and RVFW, R = 0 and 1 correspond to endocardium and epicardium, respectively. For septum, R = 0 and 1 correspond to LV endocardium and RV endocardium, respectively.

3.2 Effects of PH

From baseline to severe acute PH (PH++) and chronic PH with RVFW thickness 9 mm (WT9), peak RV pressure was increased from 28 mmHg at baseline to 76 mmHg at PH++ and to 85 mmHg at WT9 (Fig. 4a); peak LV pressure was relatively unchanged from baseline to PH++ but reduced by 46 mmHg at WT9 (Fig. 4b); mPAP was increased from 14 mmHg at baseline to 63 mmHg at PH++ and to 78 mmHg at WT9, which is consistent with the clinical definition of PH (mPAP > 20 mmHg) (Condon et al., 2019) (Fig. 4c); systolic aortic pressure was relatively unchanged from baseline to PH++ but reduced by 46 mmHg at WT9 (Fig. 4d).

In terms of cavity volumes, from baseline to severe acute PH (PH++) and chronic PH with RVFW thickness 9 mm (WT9), RV EDV increased from 99 mL at baseline to 117 mL at PH++ and to 133 mL at WT9 (Fig. 4e); RV end-systolic volume (ESV) increased from 46 mL at baseline to 63 mL at PH++ and to 110 mL at WT9 (Fig. 4e). As a result, RV EF was reduced from 54% at baseline to 45% at PH++ and to 17% at WT9 (Fig. 4e). LV EDV was relatively unchanged from baseline to PH++ but decreased from 99 mL at baseline to 86 mL at WT9 (Fig. 4f); LV ESV was relatively unchanged from baseline to PH++ but increased from 46 mL at PH++ to 59 mL at WT9 (Fig. 4f). As a result, LV EF was reduced from 54% at baseline (or PH++) to 30% at WT9 (Fig. 4f).

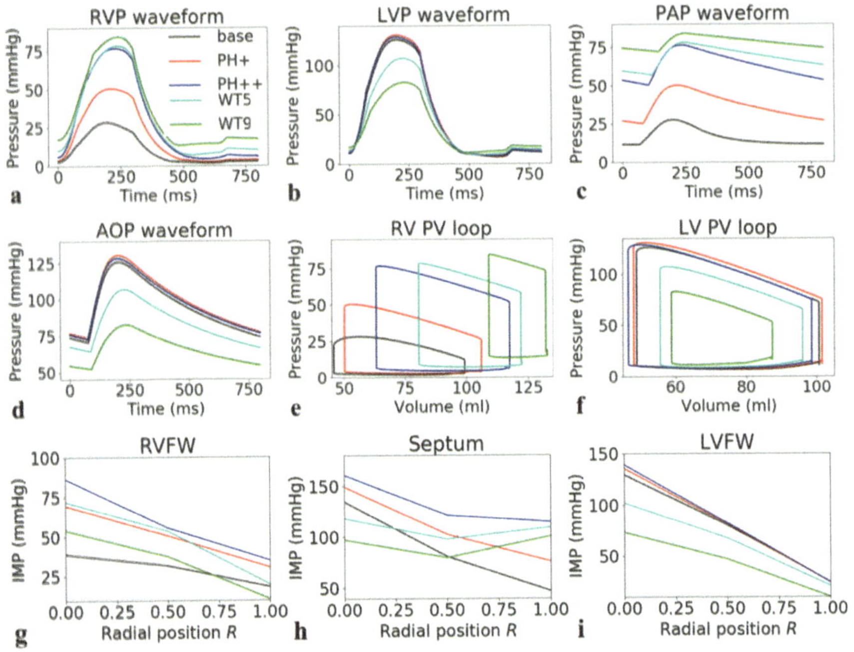

Fig. 4. Comparison of model predicted a. RV pressures, b. LV pressures, **c.** PAPs, **d.** arterial pressures, **e.** RV PV loops, **f.** LV PV loops, and transmural distribution of IMPs in the **g.** RVFW, **h.** septum and **i.** LVFW, between baseline, acute PH (PH+ and PH++) and chronic PH (WT5 and WT9). PH+ and PH++ denote moderate and severe acute PH, respectively. WT5 and WT9 denote moderate and severe chronic PH, respectively.

Intramyocardial pressure at the RVFW were all increased from baseline to WT9, especially at the endocardium (Fig. 4g), at the septum transmurally from baseline to PH++ but decreased from PH++ to WT9 (Fig. 4h), at the LVFW however, were largely unchanged from baseline to PH++ but decreased at the endocardium from PH++ to WT9 (Fig. 4i).

Total coronary flow in the RVFW is reduced from 0.25 mL to 0.22 mL at RV endocardium (~12%) but relatively unchanged at epicardium from baseline to PH++ (Fig. 5a). From PH++ to WT9, the transmural distribution of total coronary flow decreased across the RVFW and the reduction is more uniform across the wall (from 0.22 to 0.12 mL at the endocardium and from 0.27 to 0.13 mL at the epicardium) (Fig. 5a). From baseline to PH++, transmural distribution of total coronary flow in the septum is relatively unchanged at LV endocardium (R = 0) as both perfusion pressure and IMP in the LVFW are not changed whereas decreased at RV endocardium (R = 1.0) since IMP at septum increases more at RV endocardium than that at LV endocardium but perfusion pressure is unchanged (Fig. 5b). Transmural distribution of total coronary flow is reduced across the septal wall from PH++ to WT9 as the reduction of perfusion pressure is higher than that of IMP (Fig. 5b). The reduction is higher at the RV endocardium than that at the LV endocardium. From baseline to PH++, total coronary flow in the septum reduces from 0.89 mL to 0.60 mL at RV endocardium in RD circulation (33%). Total coronary flow in the LVFW is relatively unchanged transmurally from baseline to PH++ (Fig. 5c). From PH++ to WT9, transmural distribution of total coronary flow reduces in the LVFW and the reduction is more significant at the epicardium (from 0.48 to 0.28 mL) than that at the endocardium (from 0.28 to 0.21 mL) (Fig. 5c). This can be explained because in WT9, perfusion pressure is decreased and the reduction of peak IMP at the LVFW epicardium is less significant as compared to that at the endocardium, impeding more flow at the LVFW epicardium.

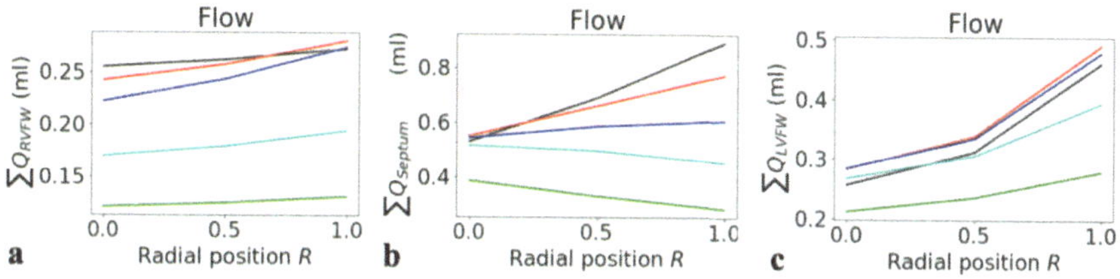

Fig. 5. Comparison of model predicted transmural distribution of coronary blood flow in **a.** RVFW, **b.** septum, and **c.** LVFW between baseline, acute PH (PH+ and PH++) and chronic PH (WT5 and WT9).

4 Discussion

The overall finding of this study is that the myocardium in both the RVFW and the septum is susceptible to ischemia in both acute and chronic PH but the mechanism that causes this hypoperfusion differs in the two conditions. Specific findings from the computational study are: **1)** IMP across the RVFW and the interventricular septum near to the RV endocardial region increased significantly in acute PH which in turn reduced coronary flow in the RVFW and septum. **2)** perfusion pressure that is associated with the aortic pressure reduced over whole cardiac cycle for each network in chronic PH whereas IMP across RVFW and septal region reduced during systole and unchanged across the LVFW, leading to reduction of coronary flow in diastole in the LVFW but in systole in the septum and RVFW. **3)** the septal region closer to the RV endocardium is more vulnerable to ischemia in both acute and chronic PH whereas RV across the FW is more vulnerable to ischemia in chronic PH. These findings suggest that coronary perfusion of the heart is affected by PH, which in turn, may play a role in the transition from compensated to decompensated PH. In clinic, the impaired right and left coronary perfusion could serve as a predictive biomarker for early diagnosis of right heart failure. Potential use of pharmacological strategies such as vasodilator may help mitigate these perfusion deficits.

Effects of PH on Biventricular Function

The elevated mPAP (63 mmHg in acute PH and 78 mmHg in chronic PH (Fig. 4c)) and peak RV pressure (77 mmHg in acute PH and 85 mmHg in chronic PH (Fig. 4a)) are consistent with previous experimental or clinical measurements. Previous experimental studies on dogs show that mPAP increased from 13 to 36 mmHg with acute PH induced by acute embolism (Bellofiore et al., 2013) and peak RV pressure increased from 34 to 47 mmHg in acute PH induced by balloon inflation (Zong et al., 2002). A previous clinical study found that mPAP is elevated at 53 mmHg in patients with PH (Spruijt et al., 2015). Other clinical studies found that for patients with PH, LV EF and RV EF were reduced to 24% (Miller et al., 2013) and 38%, respectively, which agrees with our model predictions of LV EF 32% (Pirat et al., 2006) and RV EF 17% in chronic PH. Besides ventricular and vascular pressure, the model also predicted the effects of PH on IMP that is challenging to measure in the thin RV wall. At baseline, peak IMP at the RVFW varies transmurally from 40 mmHg at the endocardium to 20 mmHg at the epicardium (Fig. 2d). Compared to the IMP at the LVFW (that varies from 128 to 30 mmHg), the gradient of RV IMP is substantially lower (Fig. 2d). The values of RV peak IMP are also broadly consistent with measurements in a previous study showing that peak IMP varies from 35 mmHg to 22 mmHg from the inner to outer layers of the RVFW (Baird et al., 1970).

Effects of PH on Coronary Perfusion

In baseline, our model predicts that the total coronary flow over one cardiac cycle is 1.03 mL in the LVFW, 2.11 mL in the septum and 0.79 mL in the RVFW (Fig. 5). These model predictions agree with clinical measurements derived from health human with right dominant circulation where the measured coronary blood flow is 1.03 mL in the LVFW, 2.01 mL in the septum and 0.80 mL in the RVFW (Sakamoto et al., 2013). Due to the elevated IMP in the RVFW, coronary flow (under fully vasodilated conditions) across the RVFW is reduced in acute PH and chronic PH (Fig. 5a). The model predictions are consistent with experimental findings that incremental balloon obstruction of the pulmonary artery led to RV dysfunction with compromised RCA flow in a canine model of acute pulmonary embolism (Taquini et al., 1960). An elevation of the RV pressure (that is associated with PH) is also shown to increase venous pressure and consequently reduce coronary flow (Watanabe et al., 1990). On the other hand, a canine study shows that RV lactate uptake (an index of RV ischemia) was increased from 1.9 to 3.0 μmol/ml/100 g in acute PH induced by balloon inflation (Zong et al., 2002). Subsequent animal experiments also show that there is a reduction in RV myocardial perfusion (Gold and Bache, 1982; Vlahakes et al., 1981) and biochemical evidence of RV ischemia (Vlahakes et al., 1981) with substantial pulmonary artery obstruction. Ischemia in the RV is also detected based on myocardial photon emission computed tomography (SPECT) images of patients suffering from chronic PH (Gómez et al., 2001). Taken together with the reduction in RV perfusion, it is likely that ischemia contributes to RV failure in acute PH. With an increase in RV wall thickness, a study shows that the endocardium to epicardium flow ratio in the LV under maximal vasodilation varies little from 0.78 to 0.75 with RV hypertrophy. This finding is comparable to the model prediction that the endocardium to epicardium flow ratio in the LVFW is 0.75 (Fig. 5b). As for regional myocardium, the model predicted that only RV perfusion is affected in acute PH but both LV and RV perfusion impaired in chronic PH (Fig. 5c). The impaired LV coronary flow in chronic PH may reduce the LV contractility based on the well-observed "perfusion-contraction matching" (Ross Jr, 1991), which will eventually leading to the LV dysfunction.

Limitations

There are some limitations associated with this simulation study. **First**, an idealized biventricular geometry is used in this study to represent that in a healthy human. Subject-specific geometries can be used to increase the model's realism. **Second,** we do not consider flow regulation in this study, which will be accounted in future.

Summary

In this study, we developed a novel computational modeling framework that couples the coronary perfusion with a biventricular FE model in a closed-loop, lumped model of the systemic and pulmonary circulations. The developed model can quantify the effects of acute and chronic PH on biventricular mechanics and coronary perfusion in each territory and transmural location of regional myocardium. Specifically, the model suggests that the coronary flow in the RV is reduced in both acute and chronic PH, whereas LV coronary flow is affected only in chronic PH. The findings suggest that chronic PH may lead to subclinical LV dysfunction and increased ischemic risk through reduced LV perfusion, which may indicate the transition from compensated to decompensated right HF.

References

Augustin, C.M., et al.: A computationally efficient physiologically comprehensive 3D–0D closed-loop model of the heart and circulation. Comput. Methods Appl. Mech. Eng. **386**, 114092 (2021). https://doi.org/10.1016/j.cma.2021.114092

Baird, R.J., Manktelow, R.T., Shah, P.A., Ameli, F.M.: Intramyocardial pressure: a study of its regional variations and its relationship to intraventricular pressure. J. Thorac. Cardiovasc. Surg. **59**, 810–823 (1970). https://doi.org/10.1016/s0022-5223(19)42402-1

Bellofiore, A., et al.: Impact of acute pulmonary embolization on arterial stiffening and right ventricular function in dogs. Ann. Biomed. Eng. **41**, 195–204 (2013). https://doi.org/10.1007/s10439-012-0635-z

Bernard, A., et al.: 3D echocardiographic reference ranges for normal left ventricular volumes and strain: results from the EACVI NORRE study. Eur. Heart J. Cardiovasc. Imaging **18**, 475–483 (2017). https://doi.org/10.1093/ehjci/jew284

Condon, D.F., Nickel, N.P., Anderson, R., Mirza, S., De, V.A., Perez, J.: The 6th world symposium on pulmonary hypertension: what's old is new [version 1; peer review: 3 approved]. F1000Research **8**, 1–8 (2019)

Fan, L.: Adaptive meshless point collocation methods: investigation and application to geometrically non-linear solid mechanics (2019)

Fan, L., et al.: Comparison of left ventricular function derived from subject-specific inverse finite element modeling based on 3D ECHO and magnetic resonance images. Bioengineering **11**, 735 (2024)

Fan, L., et al.: An in silico study of the effects of left ventricular assist device on right ventricular function and inter-ventricular interaction. Artif. Organs **47**, 1831–1847 (2023). https://doi.org/10.1111/aor.14649

Fan, L., Choy, J.S., Raissi, F., Kassab, G.S., Lee, L.C.: Optimization of cardiac resynchronization therapy based on a cardiac electromechanics-perfusion computational model. Comput. Biol. Med. **141**, 105050 (2021). https://doi.org/10.1016/j.compbiomed.2021.105050

Fan, L., Namani, R., Choy, J.S., Awakeem, Y., Kassab, G.S., Lee, L.C.: Role of coronary flow regulation and cardiac-coronary coupling in mechanical dyssynchrony associated with right ventricular pacing. Am. J. Physiol. Hear. Circ. Physiol. **320**, H1037–H1054 (2021b). https://doi.org/10.1152/ajpheart.00549.2020

Fan, L., Namani, R., Choy, J.S., Kassab, G.S., Lee, L.C.: Transmural distribution of coronary perfusion and myocardial work density due to alterations in ventricular loading, geometry and contractility. Front. Physiol. **12**, 1–20 (2021). https://doi.org/10.3389/fphys.2021.744855

Fan, L., Namani, R., Choy, S., Kassab, G.S., Lee, L.C.: Effects of mechanical dyssynchrony on coronary flow: insights from a computational model of coupled coronary perfusion with systemic circulation. Front. Physiol. **11**, 915 (2020)

Fan, L., Sun, Y., Choy, J.S., Kassab, G.S., Lee, L.C.: Mechanism of exercise intolerance in heart diseases predicted by a computer model of myocardial demand-supply feedback system. Comput. Methods Programs Biomed. **227**, 107188 (2022). https://doi.org/10.1016/j.cmpb.2022.107188

Fan, L., Wang, H., Kassab, G., Lee, L.C.: Review of cardiac – coronary interaction and insights from mathematical modeling 1–37 (2024). https://doi.org/10.1002/wsbm.1642

Finsberg, H., et al.: Computational quantification of patient-specific changes in ventricular dynamics associated with pulmonary hypertension. Am. J. Physiol. Hear. Circ. Physiol. **317**, H1363–H1375 (2019). https://doi.org/10.1152/AJPHEART.00094.2019

Gold, F.L., Bache, R.J.: Transmural right ventricular blood flow during acute pulmonary artery hypertension in the sedated dog. Evidence for subendocardial ischemia despite residual vasodilator reserve. Circ. Res. **51**, 196–204 (1982). https://doi.org/10.1161/01.RES.51.2.196

Gómez, A., et al.: Right ventricular ischemia in patients with primary pulmonary hypertension. J. Am. Coll. Cardiol. **38**, 1137–1142 (2001). https://doi.org/10.1016/S0735-1097(01)01496-6

Guccione, J.M., McCulloch, A.D., Waldman, L.K.: Passive material properties of intact ventricular myocardium determined from a cylindrical model. J. Biomech. Eng. **113**, 42–55 (1991)

Haddad, F., Hunt, S.A., Rosenthal, D.N., Murphy, D.J.: Right ventricular function in cardiovascular disease, part I: anatomy, physiology, aging, and functional assessment of the right ventricle. Circulation **117**, 1436–1448 (2008). https://doi.org/10.1161/CIRCULATIONAHA.107.653576

Hilde, J.M., et al.: Right ventricular dysfunction and remodeling in chronic obstructive pulmonary disease without pulmonary hypertension. J. Am. Coll. Cardiol. **62**, 1103–1111 (2013). https://doi.org/10.1016/j.jacc.2013.04.091

Ho, S.Y., Nihoyannopoulos, P.: Anatomy, echocardiography, and normal right ventricular dimensions. Heart **92**, 2–13 (2006). https://doi.org/10.1136/hrt.2005.077875

Hoeper, M.M., Granton, J.: Intensive care unit management of patients with severe pulmonary hypertension and right heart failure. Am. J. Respir. Crit. Care Med. **184**, 1114–1124 (2011). https://doi.org/10.1164/rccm.201104-0662CI

Kakaletsis, S., et al.: Untangling the mechanisms of pulmonary arterial hypertension-induced right ventricular stiffening in a large animal model. Acta Biomater. **171**, 155–165 (2023). https://doi.org/10.1016/j.actbio.2023.09.043

Lumens, J., Delhaas, T., Kirn, B., Arts, T.: Three-wall segment (TriSeg) model describing mechanics and hemodynamics of ventricular interaction. Ann. Biomed. Eng. **37**, 2234–2255 (2009). https://doi.org/10.1007/s10439-009-9774-2

Manohar, M., Bisgard, G.E., Bullard, V., Rankin, J.H.G.: Blood flow in the hypertrophied right ventricular myocardium of unanesthetized ponies. Am. J. Physiol. Hear. Circ. Physiol. **240**, H881–H888 (1981). https://doi.org/10.1152/ajpheart.1981.240.6.h881

McLaughlin, V.V., et al.: A report of the American College of Cardiology foundation task force on expert consensus documents and the American Heart Association. Circulation **119**, 2250–2294 (2009). https://doi.org/10.1161/CIRCULATIONAHA.109.192230

Mendiola, E.A., et al.: Right ventricular architectural remodeling and functional adaptation in pulmonary hypertension. Circ. Hear. Fail. **16**, E009768 (2023). https://doi.org/10.1161/CIRCHEARTFAILURE.122.009768

Miller, W.L., Grill, D.E., Borlaug, B.A.: Clinical features, hemodynamics, and outcomes of pulmonary hypertension due to chronic heart failure with reduced ejection fraction: pulmonary hypertension and heart failure. JACC Hear. Fail. **1**, 290–299 (2013). https://doi.org/10.1016/j.jchf.2013.05.001

Namani, R., Kassab, G.S.G.G.S., Lanir, Y.: Integrative model of coronary flow in anatomically based vasculature under myogenic, shear, and metabolic regulation. J. Gen. Physiol. **150**, 145–168 (2018). https://doi.org/10.1085/jgp.201711795

Nordsletten, D., McCormick, M., Kilner, P.J., Hunter, P.J., Kay, D., Smith, N.P.: Fluid–solid coupling for the investigation of diastolic and systolic human left ventricular function. Int. J. Numer. Methods Biomed. Eng. **27**, 1017–1039 (2011). https://doi.org/10.1002/cnm.1405

Pirat, B., McCulloch, M.L., Zoghbi, W.A.: Evaluation of global and regional right ventricular systolic function in patients with pulmonary hypertension using a novel speckle tracking method. Am. J. Cardiol. **98**, 699–704 (2006). https://doi.org/10.1016/j.amjcard.2006.03.056

Richardson, S.I.H., et al.: A poroelastic immersed finite element framework for modelling cardiac perfusion and fluid–structure interaction. Int. J. Numer. Methods Biomed. Eng. **37**, 1–33 (2021). https://doi.org/10.1002/cnm.3446.A

Ross, J., Jr.: Point of view myocardial perfusion-contraction matching implications for coronary heart disease and hibernation. Circulation **83**, 1076–1084 (1991)

Ryan, J.J., Archer, S.L.: The right ventricle in pulmonary arterial hypertension: disorders of metabolism, angiogenesis and adrenergic signaling in right ventricular failure. Circ. Res. **115**, 176–188 (2014). https://doi.org/10.1161/CIRCRESAHA.113.301129

Sakamoto, S., et al.: Relation of distribution of coronary blood flow volume to coronary artery dominance. Am. J. Cardiol. **111**, 1420–1424 (2013). https://doi.org/10.1016/j.amjcard.2013.01.290

Shavik, S.M., et al.: Computational modeling studies of the roles of left ventricular geometry, afterload, and muscle contractility on myocardial strains in heart failure with preserved ejection fraction. J. Cardiovasc. Transl. Res. **14**, 1131–1145 (2021). https://doi.org/10.1007/s12265-021-10130-y

Shavik, S.M., Zhong, L., Zhao, X., Lee, L.C.: In-silico assessment of the effects of right ventricular assist device on pulmonary arterial hypertension using an image based biventricular modeling framework. Mech. Res. Commun. **97**, 101–111 (2019). https://doi.org/10.1016/j.mechrescom.2019.04.008

Spruijt, O.A., et al.: The effects of exercise on right ventricular contractility and right ventricular-arterial coupling in pulmonary hypertension. Am. J. Respir. Crit. Care Med. **191**, 1050–1057 (2015). https://doi.org/10.1164/rccm.201412-2271OC

Streeter, D.D., Spotnitz, H.M., Patel, D.P., Ross, J., Sonnenblick, E.H.: Fiber orientation in the canine left ventricle during diastole and systole. Circ. Res. **24**, 339–347 (1969). https://doi.org/10.1161/01.RES.24.3.339

Taquini, A.C., Fermoso, J.D., Aramendia, P.: Behavior of the right ventricle following acute constriction of the pulmonary artery. Circ. Res. **8**, 315–318 (1960). https://doi.org/10.1161/01.RES.8.2.315

Vlahakes, G.J., Turley, K., Hoffman, J.I.E.: The pathophysiology of failure in acute right ventricular hypertension: hemodynamic and biochemical correlations. Circulation **63**, 87–95 (1981). https://doi.org/10.1161/01.CIR.63.1.87

Watanabe, J., Levine, M.J., Bellotto, F., Johnson, R.G., Grossman, W.: Effects of coronary venous pressure on left ventricular diastolic distensibility. Circ. Res. **67**, 923–932 (1990)

Zong, P., Tune, J.D., Setty, S., Downey, H.F.: Endogenous nitric oxide regulates right coronary blood flow during acute pulmonary hypertension in conscious dogs. Basic Res. Cardiol. **97**, 392–398 (2002). https://doi.org/10.1007/s003950200048

Cranial Implant with 3D Printing Technology and Polymethyl Methacrylate (PMMA): A Low Cost Option of Cranioplasty for Patients in Developing Country

Sambardhan Dabadi(✉) and Raju Raj Dhungel

Department of Biomedical Engineering, Annapurna Neurological Institute and Allied Sciences, Maitighar, Kathmandu, Nepal
sambardhan51@gmail.com

Abstract. Cranioplasty is one of the most common neurosurgical procedures performed to repair cranial defect. Many materials and fabrication technique are used to prepare cranial implant in cases where autologous bone is not available. Polymethyl Methacrylate (PMMA) is one of the widely chosen polymer as bone substitute, often termed as bone cement. PMMA fabricated using 3D printed models used in our study have shown better fit, symmetrical shape and restore aesthetic looks of patients. The use of 3D printed implants in medical procedures has several advantages over traditional manufacturing methods. 3D printing allows for greater precision, customization, and quicker implant time. Fabrication of custom shape and sized PMMA implants are shown to have promising results in our study. This has helped to reduce the financial burden to patients as these customized implants are way too cheaper in comparison to others materials and implant preparation. In developing and resource limited countries like Nepal, where patient have to pay out of their pocket for treatment, this 3D printing technology is proven to be a best option for cranioplasty.

Keywords: 3D printing · bone cement · cranial implants

1 Introduction

Cranioplasty is one of the common surgical procedure in neurosurgery. This sort of surgery is usually aimed to reconstruct the cranial defect. In general, defects in the skull can occur due to variety of reasons such as decompressive craniectomy followed by neurological conditions like brain stroke or traumatic brain injury, and tumor related to bone or meninges that potentially cause deformity in cranial bone [1]. Whatever be the reason for the cranial defect, these deformities are highly dangerous to underlying brain. Since, brain is a critical structure and without protection of skull, brain could easily sustain injuries. So, cranioplasty is highly recommended to provide adequate protection to delicate organ, brain [2]. On the other hand, cranioplasty is equally important to maintain the aesthetics of a person and maintain a regular shape of the skull, as even a small deformity in the skull, horribly alters the looks of a person [3].

C. T. Lim et al. (Eds.): ICBME 2024, IFMBE Proceedings 132, pp. 25–33, 2025.
https://doi.org/10.1007/978-3-032-04372-6_3

Since ages, humans have been sustaining injuries on head due to variety of reasons and are having deformities in the skull. In the past, deliberate removal of skull bone was not a common procedure, however, fall, trauma, war, physical assaults were major contributing factors for cranial defect [4]. During those era, such defects were reconstructed with either external helmets and cage like devices or some implants. Mainly such implants included metals which were inert to human body like gold or silver [5]. With time, development of neurosurgery led to surgical procedures like decompressive craniectomy to treat swelling brain and to reduce intracranial pressure, many materials were researched and experimented to be used as potential cranial implants. In present context, reconstruction of cranial defect is not just confined to protection of underlying brain but is linked with aesthetics, hence 3D designed and patient specific implants are trending for the surgery [6].

Looking back at the history of cranioplasty, various implants have been tested and used. In the early 18th century, gold and silver were among the popular choice of material for the implant. Fallopius was the first one to describe the repair of cranial defects using metallic implants. Gradually, the use of animal Ivory became popular, however, due to rejection and infection, those were soon replaced by other synthetic materials. In the early 19th Century Aluminum was introduced as it was light weighted and non-corrosive in nature. While war was at its peak and injuries were booming, different synthetic and acrylic materials were tested for cranial implants. Celluloids was one of those synthetic material trialed during that era. Later on, homologous bone grafts were used for treatment but they had higher rejection rates. Plaster of Paris was also one of the experimented materials, but was not effective enough as it was prone to resorption [7]. In the late 19th century and early 20th century, Polymethyl methacrylate was introduced and soon gained popularity as a bone cement. However, exothermic nature of PMMA and hard to remodel larger defect made it a less chosen option for cranial defect repair. Titanium mesh became popular and was widely used. Many neurosurgeons still use such mesh to repair the defect [8]. In 21st century, highly biocompatible materials were developed such as hydroxyapatite and Polyether Ether-Ketone (PEEK) [9]. Though they are highly compatible as bone substitute, higher prices of such materials have made those implants out of reach for many patients [5].

In most of the countries, insurance and health policy are not strong enough to cover up all the expenses of the patients. Since cranioplasty is usually performed after at least one or two surgical procedures, patients will have consumed all their savings in those procedures living them with limited finances to have perfect and expensive implants. In Nepal, patients are either having a private insurance or they have to pay out of their pocket for the treatment. So, in search of technology that could benefit the patients at cheaper cost, we were able to create a customized 3D designed PMMA based implant. With this, technology patients could have the perfect implant fitting their defect at relatively cheaper price as compared to modern day implants made of PEEK, titanium or hydroxyapatite [10]. Additionally, as a low-middle income country (LMIC), Nepal does not have sophisticated technology to produce such implants within the country, so such implants need to be ordered from the neighboring countries like India or China. With limited resources, the PMMA based implant is able to solve the concerns of many patients with cranial defects in Nepal, this could be a possible option in many other developing and LMICs as well [11].

2 Method of Preparation

With a motive to prepare low cost implant, we chose PMMA as a material for implant and some freeware, software freely available over the internet and a simple 3D printer. The method of designing and development were carried out within the hospital premises (Annapurna Neurological Institute and Allied Sciences) and maintaining the sterility required for making an implant.

The development process started with a computed tomography (CT) scan of a patient taken at thin slice. The parameters for the CT scan imaging is shown in Table 1. During scanning, thinner the slice of image better will be the 3D reconstruction, so, in most of our cases the slice thickness was below 1mm. After CT scan images were acquired, different software was used to read the image, reconstruct into 3D model and design the required implant. The list of software used are discussed below with their respective information.

Table 1. Parameters for CT images.

Slice thickness	Less than 1mm
Orientation	Axial
Gantry Tilt	0° (Zero Degree)
Display FoV	250
Slice Increment	≤ 1
Pixel Size	≤ 0.5
Scan FoV	Large
Slice Spacing	Contiguous Slices

Radiant DICOM viewer

Radiant DICOM viewer is a freeware, available on the internet for reading and manipulation of medical images. The DICOM file was loaded to the system and 3D reconstruction was performed (Fig. 1). The 3D tool in radiant allowed an easy and convenient way to transform the 2D images into a 3D file. The 3D obtained was then transferred to stereolithographic (STL) format, which is an editable format for 3D designing.

Meshmixer

Meshmixer is a 3D designing and editing software. This software allowed easy manipulation of our 3D model (Fig. 2). The STL file was loaded to meshmixer app and region of interest was selected. Then a layer was added to fill in the defect with the desired shape and thickness. To ensure the symmetricity, shapes were modelled based on the opposite side of the skull where the actual bone of patient was present. Usually,

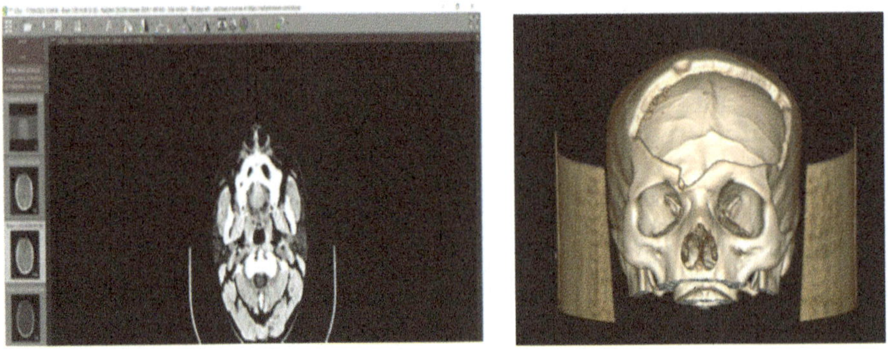

Fig. 1. 3D Image reconstruction in radiant DICOM viewer

thickness of added layer was based on the adjacent bones to the implant being designed. Once the model covered the defect, bilaterally symmetrical in outline and had enough thickness, the model was exported as STL file.

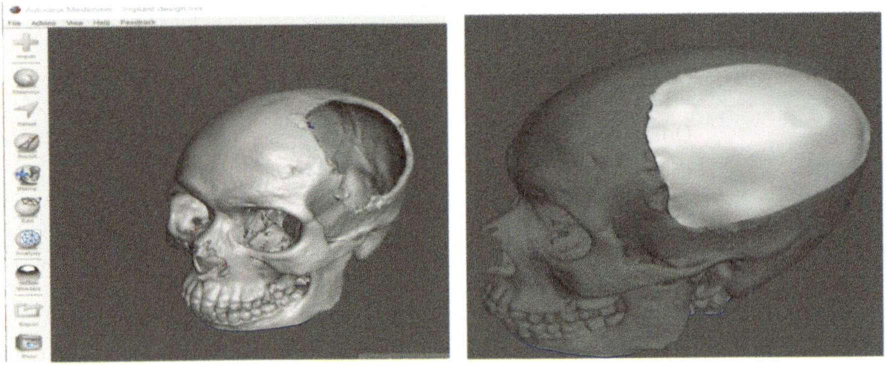

Fig. 2. 3D manipulation and designing using Meshmixer

Upstudio

Upstudio is an integrated application for the UpBox 3D printer. This application was used to print the 3D model of the implant (Fig. 3). A variety of tools are incorporated for the easy and smooth operation of the 3D printer.

Hence our production started with a CT scan image in DICOM format, which was used to generate 3D model. The model was manipulated and desired shape of implant was designed with meshmixer and the model was printed with upstudio using a upbox+ 3D printer (Fig. 4).

The printed model was made of Poly-Lactic Acid (PLA) or ABS, both being plastic material, which was not suitable for cranial implants. So, the PLA based model had to be converted to an implantable material. This was achieved with silicon casting and molding technique. The PLA model was casted in a silicon and then removed once dried.

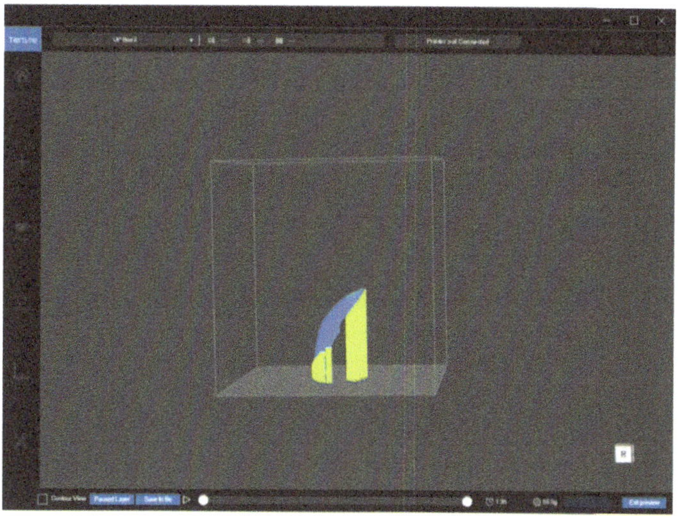

Fig. 3. Upstudio printing platform

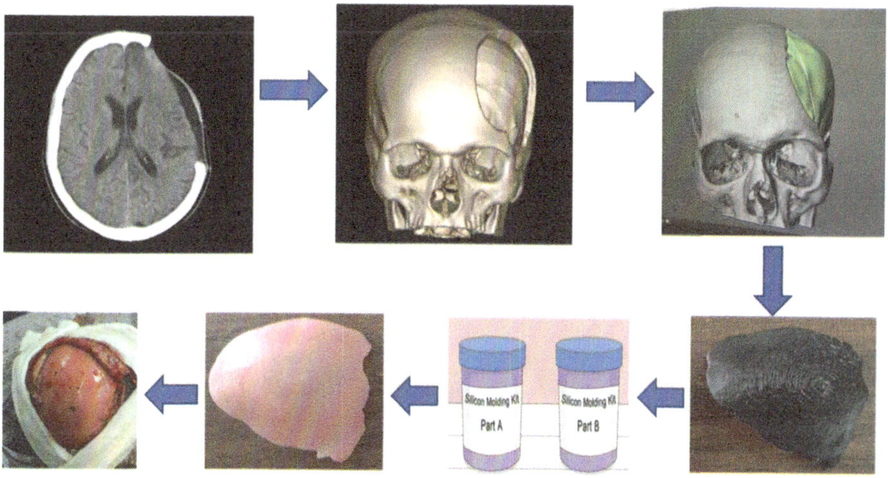

Fig. 4. Development procedure of PMMA implant

The silicon mold was then cut open to take out the PLA based model and mixture of bone cement powder with liquid methyl methacrylate (MMA) is poured into the gap. Once, it hardens the PMMA takes the shape of the model that was initially printed. Once, modelled into required shape, perforation of the model is done using a hand drill. The perforated model allows any liquid (Cerebrospinal Fluid (CSF) or blood) to leak out from the cranial case and maintain pressure inside the cranial cage. However, such perforation would not change the strength of the material.

3 Results

One hundred and eighteen cases of cranioplasty has been successfully completed using customized implants using 3D printer and PMMA. Multiple centers have already used the implant including different centers in Kathmandu, Pokhara, Biratnagar, Dhangadi and other big cities within Nepal. The classification of cases based on the site of implant is categorized in Table 2.

Table 2. Classification of cases based on surgical site

Site of Implant	Number of cases
Fronto-Temporo-parietal	42
Parietal	21
Frontal	29
Bi-Frontal/ Bilateral	26
Total	**118**

In all the operated cases, the implant was exactly fit to the cranial defect. The implant was pre-perforated for screws, this helped surgeons to implant it quickly and ensuring the critical time during the surgery is saved. (Fig. 5) The immediate post-operative CT scan of patient showed accurate fit of implant and the implant being almost symmetrical in outline, preserving the aesthetic looks of the patient. (Fig. 6(right)) Moreover, the implant did not have any cases of post-surgery complications including infection and implant rejection. Hence, in consideration to all these points, PMMA based 3D printed implant was a better option for cranioplasty.

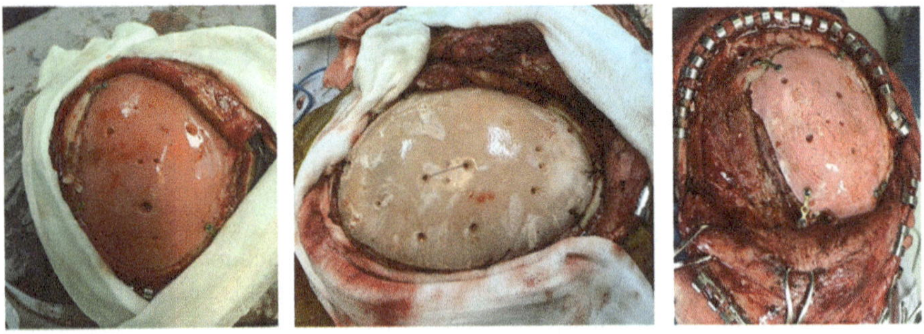

Fig. 5. Intraoperative images of some implants

Though Titanium and PEEK also have all these characteristics and features, they are quite expensive and moreover, these implants are not customizable in Nepal. The tentative cost comparison of PMMA, PEEK and Titanium is presented in Table 3.

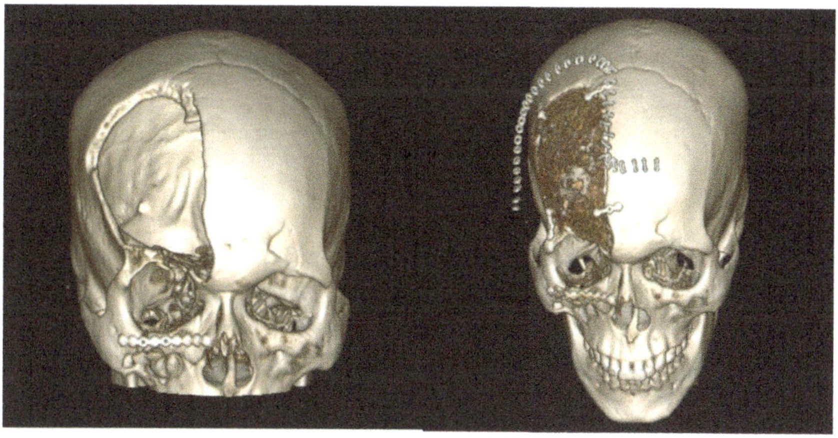

Fig. 6. Initial cranial defect (left), restored cranial shape post-operative image(right)

Table 3. Cost comparison for different implants in Nepal.

Material of Implant (Customized 3D printed and patient specific)	Tentative Cost (In Nepal)
Polymethyl Methacrylate	300 USD
Titanium	1000 USD
Polyethyl ether ketone	1500 USD

4 Discussion

Although a wide variety of implant materials are available for cranial implants, including autologous bone, animal grafts, metals, ceramics, and polymers. Some of the most commonly used implant materials are PMMA, hydroxyapatite, PEEK, high-density polyethylene (HDPE), titanium, and stainless steel [12]. All of these materials are inert, biocompatible, malleable, and provide adequate strength to protect brain tissue. Among their respective pros and cons, PEEK seems to have an edge benefit overall over these options, based on its approximate resemblance to cranial bone [13]. Although, metallic (Titanium) and ceramic (Hydroxyapatite) materials offer superior mechanical strength and osseointegration, making them preferable for long-term applications. The choice of material depends on factors like patient condition, defect size, cost, and risk of complications. Most of the metallic implants that are not customized are available as a mesh, and surgeons have to cut them into the desired shape to repair the defect [2]. On the other hand, powder-based PMMA has to be mixed and molded by the surgeons intra-operatively; in such cases, they are not homogenous in thickness and fail to maintain the symmetricity. While 3D-printed implants are sleeker, symmetrical, and have better clinical outcomes [6]. As in our 118 cases, there were no post-operative complications associated with implant rejection for at least six months to one year of the follow-up period. Table 4 demonstrates the overall comparison of features of PMMA, metallic and ceramic based cranial implants.

Table 4. Comparison of features for different materials.

Features	PMMA	Metallic	Ceramic
Strength	Moderate	High	Moderate (brittle)
Weight	Light-weight	Slightly Heavier	Light-weight
Biocompatibility	Good	Excellent	Excellent
Osseo-integration	Poor	Good	Excellent
Thermal Conductivity	Low (Good Insulation)	High (Possess risk of temperature sensitive reactions)	Low-moderate
Risk of Infection	Moderate	Low	Moderate
MRI/CT Compatibility	Excellent (Radiolucent)	Low (Causes artifacts)	Excellent (Radiolucent)

PMMA has shown promising results in developing a cheaper option for repairing the cranial defect. However, some limitations of PMMA, such as being exothermic in nature and hard to remodel larger defects, have made it a less favorable alternative among neurosurgeons in recent times [14]. Improvised techniques that incorporated 3D designing, printing, and remolding methods seemed to be a modernized and cost-effective solution for the patient in need. Generating a 3D-printed metallic implant requires high-end, dedicated, and medical-grade 3D printers, which are technically sophisticated and expensive for setup [5]. While implants developed using casting techniques can be designed and printed with simpler and cheaper home-based 3D printers capable of printing PLA, ABS, and other plastic-based materials, reducing the technical and financial burden on the institute [15]. However, the steps in developing an implant are slightly complicated with such a technique; it is relatively cheaper and more affordable for patients in low-middle-income (LMICs) and resource-constrained countries like Nepal. Thus, PMMA remains a viable option for cranial implants, especially in low-cost settings where affordability and ease of shaping are priorities.

5 Conclusion

With modernization in medical field many companies have already introduced sophisticated technologies and fabrication method of PEEK, titanium and stainless steel based customized implants, but these technologies are confined to limited nations and individuals. Though these products are more biocompatible, efficient and convenient to use, but higher cost had made these unaffordable for many of those in need. These are readily available only in those countries with facilities of good medical insurance, where patients do not have to worry about their medical expenses. But in many countries where medical insurance is not strong and patients have to pay out of their pocket for their treatment, these technologies are beyond their reach. Hence 3D printed patient specific implants made up of PMMA, is a best alternative in terms of cost, accuracy, shape, aesthetics and fit.

References

1. Reid Gooch, M., Gin, G.E., Kenning, T.J., German, J.W.: Complications of cranioplasty following decompressive craniectomy: analysis of 62 cases. Neurosurg. Focus **26**(6), 1–7 (2009). https://pubmed.ncbi.nlm.nih.gov/19485722/
2. Iaccarino, C., Kolias, A.G., Roumy, L.G., Fountas, K., Adeleye, A.O.: Cranioplasty following decompressive craniectomy. Front. Neurol. 10 (2020). https://pubmed.ncbi.nlm.nih.gov/32063880/
3. Csámer, L., et al.: Custom-made 3D printing-based cranioplasty using a silicone mould and PMMA. Sci. Rep. **13**(1) (2023). https://pubmed.ncbi.nlm.nih.gov/37491550/
4. Aydin, S., Kucukyuruk, B., Abuzayed, B., Aydin, S., Sanus, G.Z.: Cranioplasty: Review of Materials and Techniques, vol. 2, pp. 162–167. Thieme Medical Publishers (2011)
5. Bonda, D.J., Manjila, S., Selman, W.R., Dean, D.: The recent revolution in the design and manufacture of cranial implants: modern advancements and future directions. Neurosurgery **77**(5), 814–824 (2015). https://pubmed.ncbi.nlm.nih.gov/26171578/
6. Evins, A.I., et al.: On-demand intraoperative 3-dimensional printing of custom cranioplastic prostheses. Oper. Neurosurg. **15**(3), 341–349 (2018). https://pubmed.ncbi.nlm.nih.gov/29346608/
7. Sci-Hub. Materials used in cranioplasty: a history and analysis. Neurosurg. Focus **36**(4), E19 (2014). https://doi.org/10.3171/2014.2.FOCUS13561. https://sci-hub.se/https://pubmed.ncbi.nlm.nih.gov/24684331/
8. Goyal, S., Goyal, M.K.: Restoration of large cranial defect for cranioplasty with alloplastic cranial implant material: a case report. J. Ind. Prosthodont. Soc. **14**(2), 191–194 (2014)
9. Maravelakis, E., David, K., Antoniadis, A., Manios, A., Bilalis, N., Papaharilaou, Y.: Reverse engineering techniques for cranioplasty: a case study. J. Med. Eng. Technol. **32**(2), 115–121 (2008)
10. Dabadi, S., Dhungel, R.R.: Cranial defect repair with 3D designed models (2024). https://doi.org/10.1177/11795972241291777
11. Dabadi, S., Raj Dhungel, R., Dabadi, E.S.: Biomedical engineering in Nepal: opportunities and challenges. Annapurna J. Health Sci. **1**(1), 52–54 (2021)
12. Ratner, B.D., Hoffman, A.S., Schoen, F.J., Lemons, J.E.: An Introduction to materials in medicine. Biomater. Sci. 864 p. (2004)
13. Binhammer, A., Jakubowski, J., Antonyshyn, O., Binhammer, P.: Comparative cost-effectiveness of cranioplasty implants. Plast. Surg. **28**(1), 29–39 (2020)
14. Marchac, D., Greensmith, A.: Long-term experience with methylmethacrylate cranioplasty in craniofacial surgery. J. Plast. Reconstruct. Aesth. Surg. **61**(7), 744–752 (2008)
15. Kim, B.J., Hong, K.S., Park, K.J., Park, D.H., Chung, Y.G., Kang, S.H.: Customized cranioplasty implants using three-dimensional printers and polymethyl-methacrylate casting. J. Korean Neurosurg. Soc. **52**(6), 541 (2012)

Future-Ready Medical Vest: Real-Time Multi-Channel Auscultation for AI-Assisted Pulmonary Diagnostics

Hua Zhang[1](✉), Rohit Milind Neurgaonkar[1], Kay Wei Luee[1], Raj Pallapothu[2], and Chin Tuan Tan[2]

[1] Temasek Polytechnic, Tampines S(529757), Singapore
zhanghua@tp.edu.sg

[2] Bring About MI (S) Pte Ltd, (A portfolio company of Bio 9 Ventures-USA), Bedok S(470764), Singapore
raj@bio9ventures.com

Abstract. This paper presents the design and development of a future-ready medical vest for real-time, multi-channel auscultation, designed to enhance pulmonary diagnostics and enable data readiness for cloud-based AI processing. The wearable vest is equipped with an array of acoustic sensors to capture respiratory sounds from multiple sites simultaneously. The system digitizes these signals and transmits them in real-time for remote access, making it suitable for telehealth applications as well. With real-time data acquisition, signal filtering, and pre-processing, the vest ensures accurate and efficient diagnosis. Although AI-assisted analysis is not the primary focus, the system is specifically designed to enable future AI integration for advanced respiratory disease diagnostics.

Keywords: multi-channel auscultation · pulmonary diagnostics · AI-ready medical devices · wearable health technology · real-time data acquisition

1 Introduction and Background

Respiratory diseases are a significant global health issue, accounting for millions of deaths annually. Chronic obstructive pulmonary disease (COPD) and lower respiratory tract diseases rank as the fourth and fifth leading causes of death, respectively, claiming more than 5.73 million lives in 2021 alone, over 1 billion people suffer from acute or chronic respiratory conditions [1], ranging from asthma to more serious ailments like lung cancer and pneumonia. The ongoing rise in respiratory-related health complications, compounded by pollution, smoking, and urbanization, underscores the urgent need for innovative diagnostic technologies that can mitigate the impact of these diseases.

Auscultation, the practice of listening to internal body sounds such as heartbeats or lung sounds, remains one of the most common and essential diagnostic tools for assessing respiratory conditions. Typically, this is performed using a stethoscope, through which healthcare providers can detect abnormal respiratory sounds like wheezes, crackles, and

C. T. Lim et al. (Eds.): ICBME 2024, IFMBE Proceedings 132, pp. 34–53, 2025.
https://doi.org/10.1007/978-3-032-04372-6_4

stridor. This method provides crucial information about a patient's respiratory health, including the presence of diseases like pneumonia, bronchitis, or COPD.

However, traditional auscultation methods rely heavily on the clinician's skill, experience, and hearing acuity. This introduces a high degree of subjectivity and affects diagnostic accuracy, especially in settings where experienced clinicians are scarce.

Furthermore, conventional stethoscopes typically capture sounds from a single location at a time, requiring multiple assessments to obtain a comprehensive understanding of the patient's lung function. This single-point auscultation method lacks repeatability and can miss critical details about how respiratory conditions affect different parts of the lungs.

One major limitation of manual auscultation is its reliance on subjective judgment. The accuracy of the diagnosis is contingent upon the listener's ability to differentiate between normal and abnormal sounds. Given the complex nature of respiratory sounds, the likelihood of human error is significant, particularly in environments with less-experienced practitioners. Moreover, manual auscultation is time-consuming and often does not provide enough data for a comprehensive assessment of the lungs, limiting its use for continuous monitoring.

Digital stethoscopes offer a solution to these limitations by providing objective, high-fidelity recordings and, in some cases, AI-powered analysis to assist in sound interpretation. Table 1 summarizes commonly available e-stethoscopes and their key features, highlighting the advancements in digital auscultation technology.

However, there are currently no widely available commercial devices that offer automatic, multi-site auscultation capabilities.

This gap highlights a crucial need for technological solutions that can automatically capture and analyze respiratory sounds objectively, consistently, and in real-time. A device that can overcome the limitations of traditional stethoscopes, provide comprehensive lung diagnostics, and facilitate remote monitoring in telehealth scenarios is urgently needed [2–4, 6, 7].

1.1 Objectives of the Study

The primary goal of this study is to design an advanced medical vest system that addresses the shortcomings of traditional auscultation methods. This system represents a significant technological leap by automating the process of pulmonary sound detection and analysis across multiple auscultation sites simultaneously. By embedding an array of sensors at key anatomical locations on the vest, we aim to achieve real-time, multi-channel acoustic sensing of respiratory sounds. This will allow for faster, more accurate diagnosis of respiratory conditions and reduce the need for human intervention in data collection.

Additionally, the proposed vest system will digitize respiratory sound signals from all 11 channels and transmit them to a host device for analysis. This capability opens new possibilities for telehealth applications, where patients in remote areas or with limited access to healthcare can still receive timely, accurate diagnoses. The system's capacity for remote patient monitoring (RPM) is particularly important in the current healthcare landscape, where there is an increasing emphasis on delivering care outside of traditional healthcare settings.

Table 1. Key Features of Commercial Digital Stethoscope

Feature	AviceMD Wearable Stethoscope	Eko CORE 500 Digital Stethoscope	eMurmur Heart AI with Risonic Electronic Stethoscope	Thinklabs One Digital Stethoscope
Multi-Site Auscultation	**X** No - Single-point wearable stethoscope	**X** No - Single-point auscultation with digital enhancements	**X** No - Single-point auscultation	**X** No - Single-point auscultation
Real-Time Signal Processing	✔ Yes - Digital filtering for signal clarity	✔ Yes - Digital amplification, noise cancellation, and enhanced filtering	✔ Yes – AI - enhanced filtering for signal clarity	✔ Yes - Digital noise filtering
Sensor Fusion	**X** No	**X** No	**X** No	**X** No
AI-Driven Diagnosis	✔ Yes - Remote AI monitoring platform	✔ Yes - AI-powered murmur and atrial fibrillation (AFib) detection via Eko App	✔ Yes - AI-powered analysis for heart, lung, and bowel sounds	**X** No
AI Signal Enhancement	**X** No	✔ Yes - AI-powered sound enhancement	✔ Yes - AI-driven sound classification	**X** No
Remote Monitoring & Telemedicine	✔ Yes - Cloud-based monitoring platform	✔ Yes - Wireless streaming & Eko App cloud sharing	✔ Yes - Remote AI-powered analysis & telehealth integration	**X** No
Target Applications	Continuous pulmonary monitoring	Cardiovascular screening, heart & lung function monitoring	AI-powered auscultation for clinical & research use	Cardiology-focused auscultation

Furthermore, this system lays the groundwork for integrating AI-driven diagnostic tools. By providing a rich dataset of respiratory sounds, the vest system will enable future advancements in AI-based analysis, offering more sophisticated diagnostic capabilities over time. These AI tools could help detect patterns in respiratory sounds that are too subtle for human listeners, improving diagnostic accuracy even further.

1.2 Global Need for Automated Auscultation

Respiratory diseases are not confined to any one region; they represent a global health crisis that spans developed and developing countries alike. In many low-resource settings, access to trained healthcare professionals is limited, making early detection of respiratory conditions challenging. Automated auscultation systems, like the proposed vest, could help bridge this gap by enabling healthcare workers with minimal training to capture and transmit diagnostic-quality data for remote analysis by specialists. This would enhance the availability and quality of care in underserved communities.

1.3 Significance of Proposed Solution

Figure 1 shows the block diagram of proposed medical vest system.

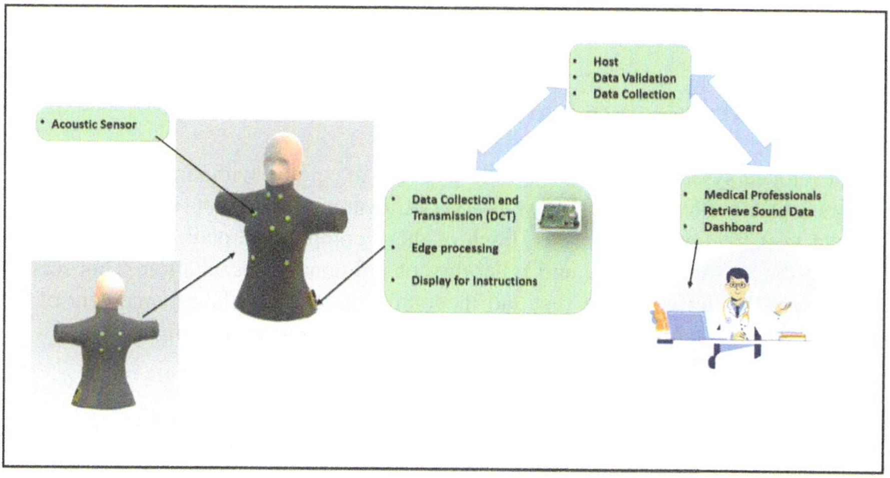

Fig. 1. Shows the block diagram of the proposed medical vest system

The proposed medical vest system offers several distinct advantages over traditional auscultation tools:

Simultaneous Multi-Site Sensing: By placing sensors at key auscultation points on the body, the vest allows for the simultaneous collection of data from multiple locations. This feature significantly improves the speed and accuracy of lung function assessments, as it provides a comprehensive view of respiratory activity across the lungs.

Telehealth and Remote Monitoring: The vest's ability to digitize and transmit data to a host device enables telehealth applications, making it a powerful tool for patients in remote areas. Real-time data transmission ensures that healthcare providers can promptly assess patient conditions, potentially preventing the escalation of treatable diseases.

Foundation for AI-Enabled Diagnostics: As AI continues to revolutionize healthcare, the vest's multi-channel data collection and synchronization provide an ideal platform for developing AI-based diagnostic algorithms. These algorithms could eventually lead to automated diagnoses of conditions such as COPD, asthma, and even more complex respiratory disorders.

Local Relevance: Singapore's National Healthcare Strategy
In Singapore, where the government's national healthcare strategy focuses on bringing care closer to the home, there is a growing demand for user-friendly devices that can facilitate remote health monitoring. The proposed vest system aligns with this initiative, offering a practical solution for monitoring pulmonary health outside of hospital settings. It could also support teleconsultations, reducing the burden on healthcare facilities and allowing for more efficient use of resources in a country where the aging population is placing increasing pressure on healthcare services.

By addressing these global and local needs, the system has the potential to improve diagnostic accuracy, reduce mortality rates from respiratory diseases, and contribute to the overall advancement of healthcare technology.

2 System Architecture and Technical Framework

The developed system consists of a wearable vest integrated with an array of acoustic sensors, a Data Collection and Transmission (DCT) unit, and a remote dashboard (RDB) for data visualization and analysis. Each of these components plays a critical role in ensuring the accuracy, speed, and reliability of pulmonary auscultation. This section will explore the design of the vest and its control electronics, the arrangement of the acoustic sensors, and the process of data acquisition and pre-processing.

2.1 Vest and Acoustic Sensor Array (VASA)

The wearable vest serves as the primary interface for patient monitoring, where multiple acoustic sensors are placed to detect respiratory sounds at pre-identified auscultation points. In this system, we employ piezoelectric sensors that offer high sensitivity and durability, making them ideal for capturing the subtle variations in respiratory sounds.

Auscultation Points: The vest is equipped with 11 strategically distributed sensors covering both the front and back of the torso to maximize the detection of abnormal respiratory sounds as shown in Fig. 2.

Front Sensors: Two sensors are positioned near the neck to capture stridor sounds, which indicate obstructions in the upper airways. Two additional sensors placed at the upper and lower lung fields on either side of the chest detect rhonchi and pleural friction rubs, respectively. These are crucial indicators of obstructions or inflammation in the lower airways.

Back Sensors: On the back, sensors are positioned to detect wheezes and crackles, which are common in conditions like asthma or pneumonia. The sensors near the costophrenic angles of the lungs (the lowest parts of the lungs) are particularly sensitive to crackles, aiding in the diagnosis of diseases like pulmonary edema or fibrosis.

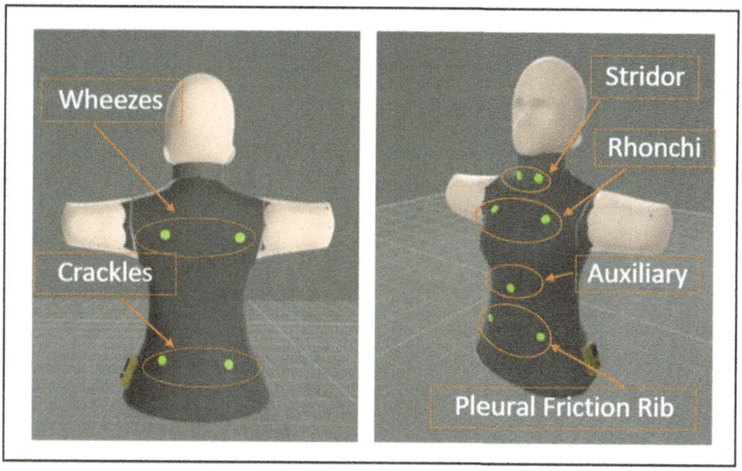

Fig. 2. Auscultation Points and Abnormal Breathing Sounds

This geographic distribution ensures that the vest can capture a wide range of respiratory sounds, each associated with specific pulmonary conditions. The real-time data collection across multiple channels allows healthcare professionals to map lung activity and locate the source of abnormal sounds with precision. This approach improves on traditional stethoscopes, which rely on subjective assessments and often miss critical sounds due to their single-point detection mechanism. The medical vest is equipped with one sensor at each designated location, covering the entire area of the lung and capable of mapping both the front and back [8].

2.2 Data Collection and Transmission (DCT) Unit

The Data Collection and Transmission (DCT) unit is the system's core for analog signal conditioning, data pre-processing, and transmission to a host device. The DCT unit manages the concurrent multi-channel acquisition of data, ensuring minimal latency and high signal integrity. Figure 3 shows the block diagram of DCT, illustrating the flow from sensor signal through filtering, ADC conversion and digital filtering.

The unit is designed to handle signals from all 11 acoustic sensors and perform initial signal conditioning, including amplification and filtering, before digitizing the signals through an Analog-to-Digital Converter (ADC) of the microcontroller unit (MCU) STM32. To ensure the signals are accurate and near noise-free, each channel undergoes band-pass filtering to isolate the frequencies relevant to respiratory sounds while suppressing unwanted noise. Main Functions of DCT include:

Hybrid Filtering: To maintain signal clarity, the DCT employs a hybrid filtering technique, ensuring the acoustic signals are captured without distortion. This is critical for accurate diagnostics, as distorted waveforms can lead to misinterpretation of respiratory conditions. The firmware allows for dynamic adjustment of cut-off frequencies for each sensor channel, tailoring the response for specific auscultation points. This will be elaborated on in the next section.

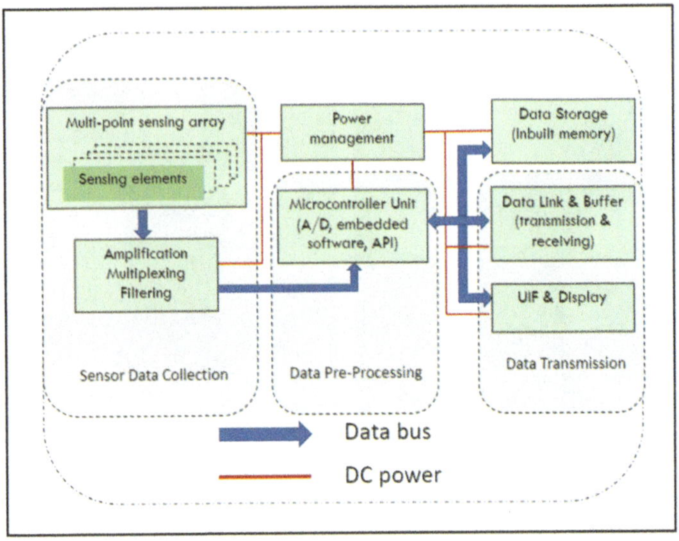

Fig. 3. Block diagram of Data Collection and Transmission (DCT) unit

High-Speed Data Acquisition: The DCT is designed for high-speed data acquisition with sampling rates exceeding 35 kHz per channel. This is essential for capturing the full detail of respiratory sounds, which can range from low-frequency murmurs to high-pitched wheezes. The multi-channel setup allows all 11 sensors to operate simultaneously, providing a comprehensive view of lung activity.

Real-Time Data Transmission: With a STM32 Virtual Com Port which is configured under USB Communication Device Class, the communication rate is high enough (up to 12 Mbps) to ensures that data from the vest is transmitted to the host in real-time. This high-speed transmission is critical for certain remote monitoring applications, where delayed or incomplete data could impact patient outcomes.

Edge Processing and AI Integration: The system incorporates edge processing capabilities, allowing the implementation of pre-trained AI models directly on the MCU. This feature enables real-time pre-assessment of the data, providing healthcare professionals with preliminary alerts when abnormal respiratory sounds are detected. By processing the data locally before transmission, the system reduces the amount of data that needs to be sent to the host, improving efficiency and reducing latency in telehealth applications.

Pre-Assessment and Telehealth Alerts: In telehealth scenarios, the system can generate real-time alerts based on the pre-assessed data, notifying medical professionals of any irregularities. This is particularly useful in remote patient monitoring (RPM) applications, where timely interventions could prevent complications or provide early diagnoses of conditions like COPD or pneumonia.

Signal Validity and Real-Time Sensor Checking: The DCT includes both hardware & firmware for real-time validation of sensor signals, ensuring that each acoustic sensor is functioning correctly. This feature is essential for maintaining the integrity of the data and preventing diagnostic errors due to faulty sensors or signal loss.

Data Buffer and Storage: The DCT is equipped with an in-built data storage buffer, allowing it to store the acquired data for up to 20 s (as optional selection) before transmitting it to the host. This feature is crucial for ensuring seamless data transmission in scenarios where network connectivity may be unstable or interrupted. The buffer ensures that no data is lost, maintaining the accuracy and continuity of the patient's respiratory record.

2.3 Firmware and System Interface

The system is controlled by an MCU STM32, which manages data acquisition, processing, and transmission. The firmware includes a user interface (UI) consisting of a display and a push button to activate the system. The UI provides real-time feedback to the user, ensuring proper operation of the vest during auscultation. The user can initiate the system, monitor its status, and view simple instructions on the display.

Additionally, the firmware handles the pre-processing of data, including filtering, signal checking, and edge-based AI processing. The digitized data is then transmitted to the host for further analysis and visualization on the remote dashboard (RDB). This seamless integration of hardware and software ensures a user-friendly experience while maintaining the system's accuracy and reliability.

3 Hybrid Filter Design and Signal Processing

In this section, we discuss the design and implementation of a hybrid filter system used for processing the respiratory sound signals in our multi-channel auscultation vest. The system employs a combination of analog and digital filtering techniques to achieve high-quality signal capture and processing, ensuring accurate and low noise data acquisition across multiple channels.

3.1 Low-Pass Analog Filtering with Digital Control

The first stage of the hybrid filter design involves analog low-pass filtering using a cascaded Sallen-Key filter architecture [4]. This design achieves a high attenuation of unwanted high-frequency noise while preserving the frequencies of interest in the respiratory sound range. Specifically, two cascaded second-order Sallen-Key low-pass filters are used in series to achieve a steep roll-off and high attenuation, approximately –40 dB, beyond the cutoff frequency. Figure 4 shows a second-order programmable low pass filter (LPF) to achieve adjustable corner frequency.

In a unity-gain configuration, the transfer function *H(s)* is given by:

$$H(s) = \frac{1}{s^2 R_1 R_2 C_1 C_2 + s(R_1 C_1 + R_2 C_1 + R_2 C_2 + 1)} \tag{1}$$

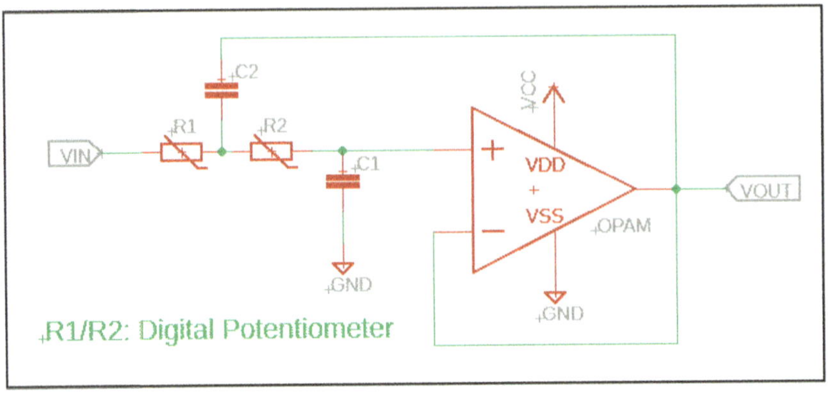

Fig. 4. Programmable Digital Low Pass Filter

With the corner frequency given by:

$$f_c = \frac{1}{2\pi\sqrt{R_1R_2C_1C_2}} \tag{2}$$

A key feature of this design is the inclusion of a digitally controlled corner frequency fc adjustment. The corner frequency of the low-pass filters can be modified in real-time through a digital potentiometer controlled by the microcontroller unit (MCU), based on the Eq. (2). This allows the system to adapt the cutoff frequency for individual auscultation channels, depending on the type of respiratory sound being captured or the region of the body being auscultated. For example, certain channels may focus on capturing high-frequency wheezing sounds, while others may target lower-frequency crackles or rhonchi.

The ability to dynamically adjust the corner frequency enhances the versatility of the system, allowing for optimal signal capture across different conditions and patient profiles. Digital High-Pass Filtering in the MCU.

In addition to the analog low-pass filtering, a digital high-pass filter (HPF) is implemented after the analog-to-digital conversion (ADC) stage. The use of a digital HPF allows us to eliminate low-frequency noise components, such as those arising from motion artifacts or environmental interference, without the need for additional physical filters in the signal chain.

The high-pass filter is implemented using an infinite impulse response (IIR) filter, which is computationally efficient and well-suited for real-time digital signal processing (DSP) on the STM32 MCU. The IIR filter is designed to have a cutoff frequency of 15 Hz, which is appropriate for removing unwanted low-frequency noise while preserving the essential components of respiratory sounds, which typically occur in higher frequency ranges. The cutoff is programmable with the digital implementation.

The filter is applied directly to the digitized signal after it is sampled at a rate of 35 kHz, with a resolution of 12 bits per sample. Given the high sampling rate, the IIR filter ensures that low-frequency components are effectively removed without introducing significant phase distortion or signal degradation.

3.2 IIR Filter Implementation

The IIR high-pass filter is designed using a second-order Butterworth filter, which provides a smooth frequency response and minimal phase distortion. The filter coefficients are determined using standard digital filter design techniques and are embedded directly into the firmware for real-time processing [4, 5, 10].

The transfer function of the IIR filter is given by:

$$y[n] = b_0 x[n] + b_1 x[n-1] + b_2 x[n-2] - a_1 y[n-1] - a_2 y[n-2] \tag{3}$$

where:

$x[n]$ is the current input sample.
$x[n-1]$ and $x[n-2]$ are the previous input samples.
$y[n]$ is the current output sample.
$y[n-1]$ and $y[n-2]$ are the previous output samples
b and a are the filter coefficients, which control the behavior of the filter.

Given the sampling rate is 35kHz and the cutoff frequency of 15Hz, the coefficients are chosen as:

$$b_0 = 0.9999; b_1 = -1.9998; b_2 = 0.9999; a_1 = -1.9998; a_2 = 0.9999$$

The STM32 MCU, which supports DSP extensions, can process this IIR filter in real time for all 11 channels.

The filter is applied in real-time to each incoming sample, ensuring that the data remains clean and ready for further processing or transmission.

3.3 Real-Time Processing and Storage

The discrete form of the transfer function, as represented by the Eq. (3), is highly suitable for real-time implementations with STM32, especially at a fixed sampling rate. This setup is efficiently managed using TIM-based interrupts, ensuring each calculation aligns precisely with incoming data intervals.

The recursive nature of this equation offers a significant advantage: it requires only a small number of memory locations, even when handling 11 channels, as it depends solely on the current and recent past inputs and outputs. This memory efficiency allows the STM32 to process data on-the-fly with minimized latency, handling each new data sample as it arrives in real time and ensure the respiratory sound signals are ready for immediate transmission to the host.

In scenarios where real-time transmission is constrained by limited network bandwidth, the system stores the latest filtered data in an SPI-based memory module (S25FL). This memory is specifically configured to hold up to 20 s of high-speed data acquisition across all 11 channels, providing a buffer for deferred processing and analysis when immediate transmission is not feasible. This approach ensures that critical data is not lost due to transmission delays, allowing for post-processing of the stored data once network conditions improve. Consequently, this setup enhances data integrity and continuity, ensuring reliable performance in bandwidth-limited environments. Figure 5 illustrates

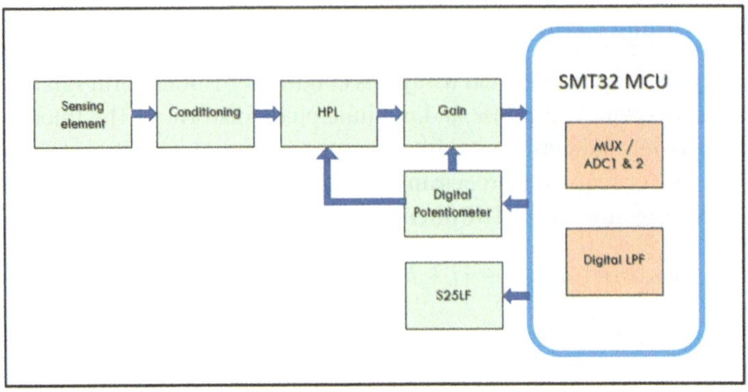

Fig. 5. Sub-System Handling Real-Time Data Capturing, Filtering, and Storage

the block diagram of the sub-system handling real-time capturing, filtering, and data storage.

Overall, the combination of analog low-pass filtering, digital high-pass filtering, and real-time processing provides a robust solution for capturing and processing respiratory sound signals. The hybrid approach ensures that the system can handle a wide range of sound frequencies, maintain signal integrity, and adapt to different clinical requirements.

4 Firmware Development and Data Transmission

The firmware is designed to enable high-speed, real-time data acquisition and filtering from 11 sensors, using dual ADCs on the STM32 microcontroller. Each ADC handles a subset of channels in parallel, with efficient Direct Memory Access (DMA) and TIM-based sampling to ensure precise data acquisition.

4.1 Real-Time Data Acquisition and Processing

To achieve continuous multi-channel data acquisition, the firmware leverages the two ADCs available on the STM32 microcontroller, each configured to handle a specific set of channels. ADC1 manages six channels, while ADC2 is responsible for the remaining five. Both ADCs operate simultaneously and continuously, ensuring no data is missed from any of the 11 channels.

Sampling Rate Control: A dedicated timer controls the sampling rate of each ADC, ensuring consistent and precise timing. This timer is configured to trigger the ADCs at a regular interval. The timer's callback function reads data from the ADCs' respective DMA buffers, which are organized as circular buffers.

DMA Configuration: Each ADC is configured with its own DMA channel to handle the transfer of sampled data into separate circular buffers. The use of DMA minimizes CPU overhead, as data is automatically moved from the ADCs to memory without CPU intervention. This configuration allows for high-speed data capture, ensuring that the system can keep up with the high sampling rate required for multi-channel auscultation.

Circular Buffer Structure: The circular buffers associated with each DMA channel continuously store the latest samples for each ADC, effectively providing a rolling window of data. This structure allows the system to retain the most recent data, which is critical for applications requiring real-time analysis.

4.2 Edge Processing and AI-Readiness

Once data is captured in the circular buffers, it undergoes real-time filtering before being transmitted or stored. The firmware implements a low-pass IIR filter algorithm as described in Sect. 4, which removes high-frequency noise while preserving the relevant respiratory sounds. This filtering occurs within the timer callback function to ensure that each sample is processed as soon as it's available, allowing for efficient real-time filtering across all 11 channels.

4.3 Edge Processing

Edge Processing and Data Pre-Processing: Edge processing is a crucial feature of the proposed auscultation system, allowing the device to perform initial data analysis locally (on the STM32 microcontroller within the Data Collection and Transmission, or DCT unit) without relying solely on cloud computing. This enables immediate diagnostic feedback based on a pre-trained AI model, significantly reducing latency and providing actionable insights to healthcare professionals in real time. The edge processing pipeline involves local data pre-processing, application of a pre-trained AI model, and integration with cloud-based machine learning for continuous model enhancement.

The edge computing workflow deployed on the STM32 MCU follows these main steps:

Local Signal Pre-Processing: The pre-processed, or filtered data is prepared for (1) the real-time calculation based on the AI model embedded on the STM32, and (2) data transmission to a host device.

Application of Pre-Trained AI Model on the Edge: The core of the edge processing is a *pre-trained AI model* that operates directly on the STM32 MCU. The AI model, which is the focus of a separate related ongoing project, is designed to recognize specific respiratory sound patterns, delivering diagnostic information for known respiratory conditions.

For the initial implementation of the current design, we utilized an existing convolutional neural network (CNN) model. Once the final AI model is fully trained and available, it will be deployed through firmware upgrading for production use.

Optimization and Implementation for STM32: To ensure computational efficiency on the STM32, fixed-point arithmetic operations are employed, using 8-bit integer quantization to enable real-time inference. This optimization balances resource constraints and performance demands.

FFT and Hamming Windows Coefficients: Given the limited hardware resources of the ARM Cortex-M4, a 256-point FFT is employed to achieve high-frequency resolution, enabling the capture of fine-grained frequency components in respiratory sounds.

FFT calculations leverage CMSIS-DSP functions to generate spectrograms, a time-frequency representation of the signals. To further optimize performance, the Hamming window coefficients are pre-calculated offline and stored as a lookup table on the STM32, minimizing runtime computation.

AI Model Deployment and Integration: For testing, the mock pre-trained model is converted to TensorFlow Lite format (.tflite). Quantization is applied to reduce model size and improve inference speed. With the help of STM32Cube.AI, the model is integrated with the STM32 firmware.

4.4 Cloud-Based AI Model Training and Continuous Improvement

While edge processing enables real-time diagnostics, the AI model is periodically retrained in the cloud to continually enhance its performance.

Data Transmission to the Cloud: With the initial edge processing, respiratory sound data is also prepared for transmission. The system utilizes serial communication for streamlined data transfer to the host. To prevent data loss due to miscommunication, connection breaks, or low-speed transfer rates, the latest data is also stored in built-in cache memory (S25LF512), configured as a circular buffer that holds up to 20 s of recent data.

Note that each new data sample for the AI model should be labeled and assessed by healthcare professionals, creating a continuously growing dataset of accurately labeled samples. This evolving dataset forms the foundation for ongoing improvements in the AI model's diagnostic accuracy.

Model Redeployment to the Edge: Following each retraining cycle, the cloud platform distills the optimized AI model for re-deployment to the STM32. This redeployment process is integrated into the STM32 initialization tasks, ensuring that polynomial coefficients are updated as they become available.

The integration of edge and cloud-based processing establishes a feedback loop, enabling the vest to continually enhance its diagnostic accuracy. This creates a self-evolving tool capable of adapting to evolving clinical needs and requirements [6, 7, 10].

5 Preliminary Device Design Validation

The initial validation of the medical vest was conducted, focusing on its signal conditioning, data requisition with high resolution at high speed, and multi-channel capability. Figure 6 illustrates the vest system, which includes integrated sensors and DCT units.

A mock AI model (preliminary training) was employed to evaluate the reliability and consistency of the data collected.

5.1 Validation of Device Design

The primary objective of this validation phase was to ensure that the device could capture high-quality respiratory sound data across multiple channels with minimal noise and artifacts, and that the pre-processed data retained diagnostic value.

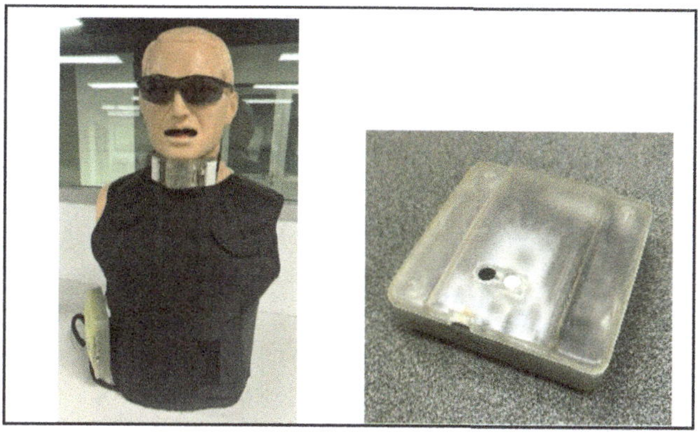

Fig. 6. Multi-Channel Auscultation Vest and DCT Unit

Methodology: The respiratory signal bandwidth was tested to confirm that it met the targeted maximum frequency range and featured firmware-controlled adjustable corner frequencies. These tests were conducted offline using wavelet FFT analysis. A mock pre-trained convolutional neural network (CNN) was trained on a labeled dataset of respiratory sounds, including samples of normal patterns (e.g., breath sounds) and abnormal patterns (e.g., wheezes, stridor) [5]. The pre-trained mock AI model (preliminary training) was then deployed to analyze the data collected by the vest, providing outputs such as spectrograms and demonstrating its ability to handle multi-channel data capture and simultaneous signal processing.

Signal Accuracy: To assess signal accuracy, an FDA-approved electronic stethoscope is used to obtain reference signals. Simultaneously, signals from the vest are recorded. Both sets of signals are sampled synchronously at a rate of 35 kHz and converted to digital format. Prior to offline analysis, the signals are normalized using a unit vector. The cross-correlation of signals from both devices is then computed to validate the accuracy of the signal from the vest.

Signal-to-Noise Ratio (SNR): The device's ability to minimize ambient noise while capturing respiratory sounds was assessed through signal-to-noise ratio (SNR) analysis. SNR was calculated as the ratio of signal power to noise power, serving as a key indicator of the vest's noise reduction performance.

Channel Consistency: The uniformity of data quality across the vest's 11 channels is checked to ensure that each sensor location was functioning effectively and capturing relevant sound data.

5.2 Results and Observations

Figure 7 depicts the bandwidth test results for a typical channel with a firmware controlled second-order low-pass filter (LPF) with two configurations set by software (STM32 firmware):

1. Gain (G1): 2.5, Cutoff Frequency (F_c): 2.8 kHz
2. Gain (G2): 1.0, Cutoff Frequency (F_c): 1.3 kHz

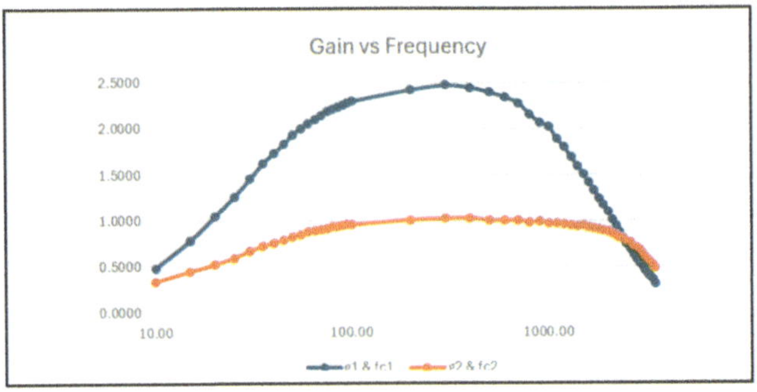

Fig. 7. Bandwidth for 2 different settings of gains & cut-off frequencies

Actual data was gathered from over 40 individuals, both healthy individuals and patients, albeit in limited quantities due to project constraints and focus. The mock CNN was evaluated with a learning rate of 0.0001, 100–500 epochs, and a batch size of 32–64. These results validate the flexibility and precision of the signal conditioning circuit for individual filters. The preliminary findings confirmed the vest's consistency in capturing respiratory sounds across multiple channels, underscoring its potential as a reliable tool for multi-channel auscultation.

Figure 8 illustrates the vest system in use during hospital-based data collection.

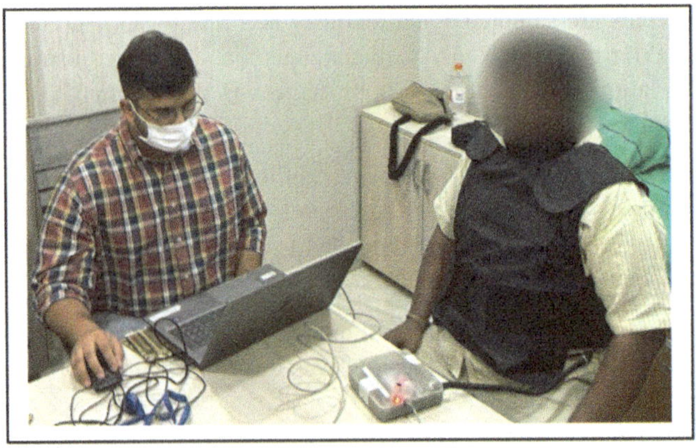

Fig. 8. Data collection at Hospital

Figure 9 presents the respiratory sound waveforms recorded from the 11 auscultation points using the proposed vest system. Note that the graph in the top-left corner represents the spare channel, which is not in use.

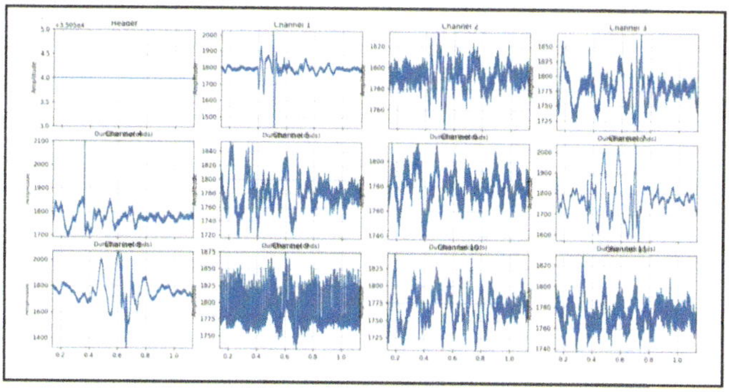

Fig. 9. Simultaneous recording of waveforms from 11 auscultation points

The traditional gold-standard tool for lung sound auscultation is the stethoscope. To enable a more objective and quantitative comparison, an FDA-approved electronic stethoscope is used. The signal-to-noise ratio (SNR) is tested and calculated for individual frequencies within the 20 Hz to 3.5 kHz range, which covers the auscultation sound spectrum relevant to clinically critical lung diseases. Tests were conducted using both a commercially available FDA-approved stethoscope and the proposed vest system to (1) verify that the proposed design achieves a high SNR across the required auscultation frequency range, (2) Compare the SNR performance of the proposed design with that of the commercial stethoscope.

As an illustration of this comparison, one representative channel from the multi-channel waveform of the vest was selected for direct comparison with the commercial e-stethoscope. Both systems were tested using signals of the same frequencies, and the SNR was computed. The results, presented in Table 2 and Fig. 10, demonstrate that the vest system has good SNR for the designed frequency range, providing a clean, high-fidelity auscultation signals, when compared to the existing product.

More comprehensive tests, including a full clinical validation study with sensitivity and specificity evaluation, will be conducted once the AI model for assisted diagnosis is fully trained under the separate project.

Furthermore, additional verification tests were performed using multi-location lung sound recording, where signals were collected from different vest positions, (1) breathing sound from front central position, and (2) coughing sound from upper left, lower and upper positions of the vest.

For these tests, Pearson correlation analysis was performed to quantify the similarity between the captured signals from the vest and the e-stethoscope. The results showed strong correlation (0.81) for breathing signal (20 s, as illustrated in Fig. 11), and perfect correlation (1.0) for all coughing signals across different positions tested.

These findings confirm that the vest system successfully captures lung sounds with minimal distortion, validating the accuracy of its electronics. While this analysis demonstrates signal integrity, future work will focus on clinical validation by correlating vest-recorded signals with clinician-labeled auscultation findings to assess diagnostic accuracy.

Table 2. Comparison of SNR between Proposed Vest System & Stethoscope

Signal frequency (Hz)	Proposed Vest System (dB)	e-Stethoscope* (dB)
20	**29.23**	**37.62**
100	**37.23**	**46.15**
300	**37.83**	**37.25**
500	**37.67**	**39.23**
800	**37.2**	**31.44**
1000	**36.78**	**26.12**
1500	**35.53**	**17.82**
3500	**24.93**	**4.2**

: FDA approved e-stethoscope

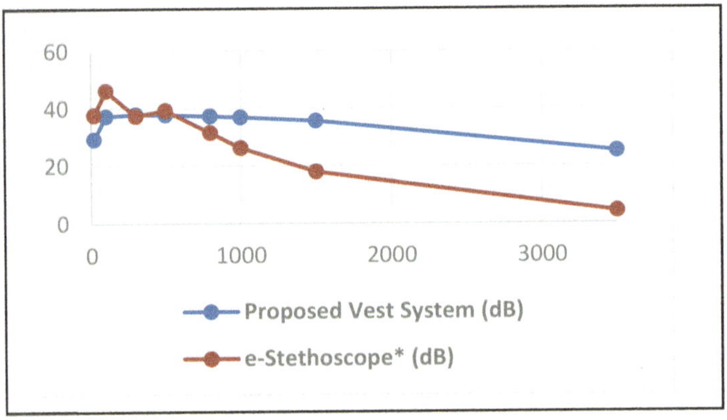

Fig. 10. Graphical Illustration of the SNR Comparison

6 Discussion, Conclusion and Future Work

6.1 Discussion

The preliminary validation confirms the medical vest meets its design objectives by capturing respiratory sound data across 11 channels within the required frequency range, with no latency or delay. Its robust hardware design ensures good noise immunity and consistent signal quality, leveraging dual ADCs, timer-controlled sampling, efficient DMA, and circular buffering.

While the validation focused on hardware and signal acquisition, further efforts are needed to harness the vest's potential for AI-assisted diagnostics. The results from the preliminary validation of the medical vest demonstrate that the device design meets the intended purpose. Specifically, the vest successfully captures respiratory sound data across all 11 channels within the required frequency range and performs simultaneous multi-channel acquisition without delay or latency. This performance aligns with the

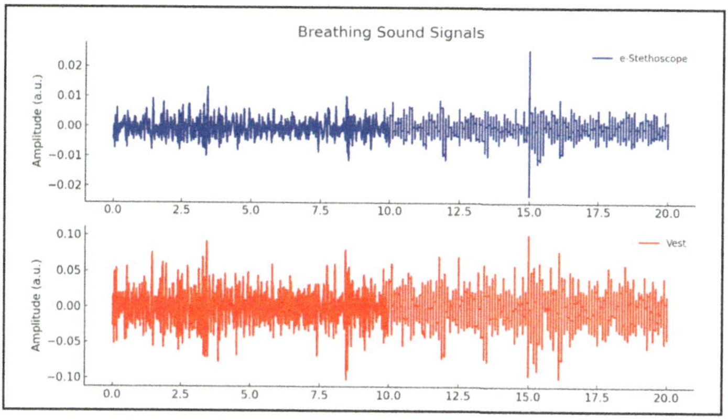

Fig. 11. Breathing sound signals from e-Stethoscope & vest system

specifications necessary for a reliable auscultation device, ensuring that each channel records sound data consistently and accurately.

The robustness of the hardware design is also evident from uniformity in signal quality across channels. The use of dual ADCs with timer-controlled sampling, supported by efficient DMA and circular buffering, has proven effective in delivering consistent data flow and managing high sampling rates. These features collectively enable the vest to provide comprehensive lung sound coverage, meeting clinical standards for auscultation.

The problem of selecting or enhancing signals from the sensor array, especially in the context of respiratory signals from simultaneous auscultation points has two practical solutions:

1. Choosing the best signal with the highest SNR from the sensor array, or
2. Enhancing signals using all signals with weighted averaging.

Each has its own advantages over the other. Solution 1 is deployed in the current design.

6.2 Conclusion

The medical vest has proven effective as a multi-channel, real-time auscultation device, delivering robust data acquisition and maintaining high signal quality. This validation underscores its readiness for future applications in remote monitoring and telehealth. The system's design effectively minimizes noise while preserving signal integrity, establishing a foundation for advanced respiratory diagnostics and AI-based analysis.

6.3 Challenges and Future Work

Considering that the model adaptation for the diverse patient profiles, the AI model requires further training on a broader dataset to account for variability in respiratory sounds across different demographics, influenced by factors such as age, body composition, and specific conditions.

Future efforts will include integrating the vest with WheezeX, a digital biomarker platform for early detection of respiratory diseases and personalized therapy. The platform serves as a centralized repository containing multi-modal patient data, including pulmonary soundscapes and other relevant clinical information, allowing the AI system to learn from diverse sources.

Figure 12 shows the block diagram of the AI-driven diagnosis vest system, which is part of a related ongoing project. The proposed system will consist of three core subsystems: (1) multi-channel auscultation vest, (2) Medical professional host with additional multi-modal medical data inputs, and (3) Cloud platform for AI model training, inference, and continuous learning.

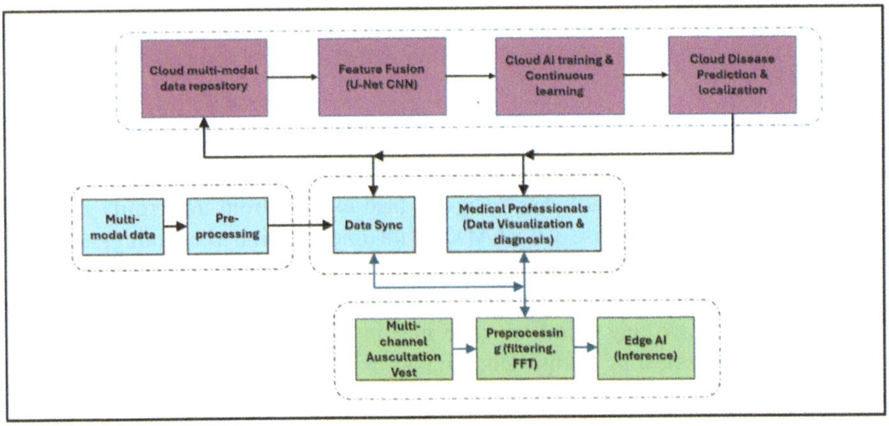

Fig. 12. Functional block of AI-driven diagnosis system.

The proposed AI system will be designed to retrieve, store, and update patient data in the platform, facilitating continuous cloud-based AI training while enabling real-time AI inference on the vest which is equipped with an edge device.

For lung disease classification, we plan to employ a U-Net CNN architecture, which will be trained on both pulmonary soundscapes and multi-modal medical data. Given that our dataset is expected to initially contain 200 patients, with a target to more than 1000 in later stages, we recognize that this may be insufficient for training a deep learning model from scratch due to the complexity of pulmonary disease classification. This presents a significant challenge for the project team.

Based on the previous data collection, it was observed that each patient's data (multi-site, 20- second recordings) can be segmented into multiple smaller samples (e.g., different auscultation positions, time frames, or frequency bands), effectively increasing the number of training samples. While this segmentation expands the dataset, the challenge is that the correlation between these segments, which must be carefully managed to prevent overfitting.

To further address data limitations, we plan to leverage a pretraining-finetuning approach, where U-Net will be first be pretrained on large open-source datasets to ensure it learns generalized pulmonary features before fine-tuning on our dataset, thereby improving diagnostic accuracy even with a relatively small patient cohort.

Additionally, future work will explore advanced data augmentation techniques to enhance the model robustness and improve generalization across different patients' profiles. These strategies will support the vest's ability to continuously improve its diagnosis accuracy through AI model refinement.

Planned validation trials will expand testing to diverse clinical environments and patient demographics, ensuring diagnostic accuracy and adaptability. These developments position the vest as a future-ready tool for clinical and telehealth applications.

Acknowledgment. The authors would like to acknowledge the support provided by the Temasek Polytechnic Research Fund under the 14th TPRF Grant Call, for funding the project titled "Development of a Medical Vest for Automated Simultaneous Multiple-Point Auscultation." This research has also benefited from the collaboration with Bring About MI (S) Pte Ltd, a portfolio company of Bio 9 Ventures-USA, whose industry expertise and resources have been invaluable in advancing the development of this medical vest. We express our sincere gratitude to all parties involved for their support and contributions.

References

1. World Health Organization. Global Health Estimates: Leading causes of death and disability (2021). https://www.who.int/data/gho/data/themes/mortality-and-global-health-estimates
2. Pasterkamp, H., Kraman, S.S., Wodicka, G.R.: Respiratory sounds: advances beyond the stethoscope. Am. J. Respir. Crit. Care Med. **156**(3), 974–987 (1997)
3. Murphy, R.L.H.: Auscultation of the respiratory system: advances in acoustic analysis. Respir. Med. **102**(7), 976–987 (2008)
4. Reichert, S., Gass, R., Brandt, C., Andrés, E.: Analysis of respiratory sounds: state of the art. Clin. Med. Circulatory Respir. Pul. Med. **2**, 45–58 (2008)
5. Wilson, D., Smith, P., Doe, J., Lee, M.: Challenges in multi-channel auscultation system design for pulmonary diagnostics. IEEE Trans. Biomed. Eng. **65**(4), 847–857 (2018)
6. Kumar, A., Sharma, K.: Telemedicine and wearable devices: a new approach in healthcare innovation. J. Med. Internet Res. **22**(9), e20012 (2020)
7. Rocha, B.M., Silva, C., Patel, R.: An open-access database for the evaluation of respiratory sound classification algorithms. Physiol. Meas. **40**(3), 035001 (2019)
8. Beňačka, R.: Auscultation of lung sounds and murmurs. Pavol Jozef Šafárik University (2020). https://patfyz.medic.upjs.sk/simulatorvzorky/Respiratoryauscultation.htm
9. Kim, Y., Hyon, Y., Lee, S., Woo, S.D., Ha, T., Chung, C.: The coming era of a new auscultation system for analyzing respiratory sounds. BMC Pulm. Med. **22**(1), 119 (2022)
10. Cohen, A., Brown, T., White, P.: Signal insights for enhancing respiratory sound analysis. IEEE Signal Process. Mag. **38**(6), 80–96 (2021)
11. Horowitz, P., Hill, W.: The Art of Electronics, 3rd edn. Cambridge University Press, Cambridge, UK (2015)
12. Petmezas, G., Rossi, F., Zhang, L.: Automated lung sound classification using a hybrid CNN-LSTM network and focal loss function. Sensors (Basel, Switzerland) **22**(3), 1232 (2022)

Estimation of Space and Time Evolution of Mg^{2+} Concentration Near a Mg^{2+}-Releasing Implant

Sandor Nemeth(✉) and Clarice Loke

Agency for Science, Technology and Research (A*STAR), Singapore Institute of Manufacturing Technology (SIMTech), Singapore, Republic of Singapore
{sandorn,phloke}@simtech.a-star.edu.sg

Abstract. According to literature reports based on *in vitro* tests, transient increase in Mg^{2+} concentration to 2 to 10 mM can promote osteogenesis around implants. However, it is not well established how the Mg^{2+} concentration may change in space and time around a Mg^{2+}-releasing implant *in vivo*, and whether therapeutic levels of Mg^{2+} can be maintained in the surrounding tissues. Thus, the temporal and spatial evolution of Mg^{2+} concentration was investigated from a $MgSO_4{\cdot}7H_2O$ containing polycaprolactone implant by numerical analysis using a rudimentary diffusion model. Diffusion coefficient of Mg^{2+} in polycaprolactone was determined experimentally, while it was estimated in body fluids and tissue from literature data. The results revealed that Mg^{2+} levels generally increase modestly and reach the therapeutic level for less than one day in this system due to fast clearing of Mg^{2+} from the tissues. Toxic levels of Mg^{2+} concentration may develop, but only for a few minutes, and can be mitigated by a thin layer of pure polymer applied on the implant.

Keywords: Polycaprolactone · Implant · Magnesium release · Osteogenesis

1 Introduction

Magnesium is an essential element in living organisms, and it is known to be involved in a large number of biochemical pathways [1]. One important role of Mg^{2+} is found in osteogenesis. It has been reported that elevated Mg^{2+} levels can accelerate osteogenesis and angiogenesis via varied mechanism if the magnesium concentration is well above the physiological level (0.65 mM–1.05 mM), somewhere in the optimal range of 2 mM to 10 mM as determined by *in vitro* experiments [2]. To take advantage of these positive effects of magnesium in regenerative medicine with regards to bone injuries, a suitable approach is to implant a temporary scaffold that can release Mg^{2+} at the site of implantation. This is to increase the magnesium concentration locally for a suitable duration and guide accelerated bone development. However, the duration of elevated magnesium level should not be excessively long as high concentration of magnesium can inhibit the mineralization process in the later stages of bone development.

The *in vitro* experiments reported in the literature are static with regards to the magnesium concentration, and it is an open question how these results may be used

C. T. Lim et al. (Eds.): ICBME 2024, IFMBE Proceedings 132, pp. 54–68, 2025.
https://doi.org/10.1007/978-3-032-04372-6_5

in vivo where the interplay between the release rate of Mg^{2+} from the scaffold, and its clearing from the vicinity of the implant creates a dynamic concentration profile [2]. Although this concentration profile could potentially be evaluated via experimental methods, the method would be complicated and difficult to plan and implement. Thus, the authors conducted numerical simulation studies as a first step to evaluate the likely fate of magnesium that is released from an implant material. A rudimentary model was used in the calculations with several simplifications that may preclude highly reliable quantitative results, but it is suitable to evaluate the main factors responsible for the release profile that one may expect. Thus, the outcome from the calculations could be used to guide the design of more sophisticated calculations and experiments while providing basic insights into magnesium release profile from an implant. The study focused on the release of magnesium sulphate from polycaprolactone (PCL) carrier since PCL is used as a biodegradable temporary implant for bone regeneration.

2 Model Setup

2.1 Geometry

A very simple release geometry was used as shown in Fig. 1 consisting of a sheet of PCL loaded with $MgSO_4{\cdot}7H_2O$, an optional layer of Mg^{2+}-free PCL, body fluid layer, tissue layer and a magnesium sink (microcapillaries). The thickness of the PCL and body fluid layers were selected arbitrarily while the tissue layer was selected based on capillary density in living tissue (see Table 1). For viability, cells can be only from a limited distance from capillaries. This can be as low as about 15 μm in mouse brain tissue [3] or as much as 67.2 ± 25.3 μm capillary-free zone in human retina [4]. Thus, the tissue layer thickness was varied in the comparable 20 μm to 80 μm range. The model was simplified by assuming the tissue to be a uniform and homogeneous layer (no cellular structure) for the purpose of magnesium transport.

All the layers were divided into 1 μm thick sub layers for the purpose of calculating changes in magnesium concentration.

2.2 Initial Mg^{2+} Concentrations

The concentration of $MgSO_4{\cdot}7H_2O$ in the PCL implant material was selected at two levels, 480 mM and 990 mM, approximately corresponding 10 w% and 20 w% salt in the polymer composite. Further, it was assumed that the salt is distributed homogeneously in the implant. When a pure PCL coating layer was considered on the implant, it had 0 mM initial salt content. The body fluid and tissue had 1 mM initial magnesium concentration, while the capillaries always had a fixed 1 mM magnesium concentration.

2.3 Diffusion Coefficients

The numerical model relied solely on simple diffusional transport of magnesium from the implant to the capillaries. The numerical value of diffusion coefficient of $MgSO_4$ in the presence of water was determined in PCL experimentally and described in a conference

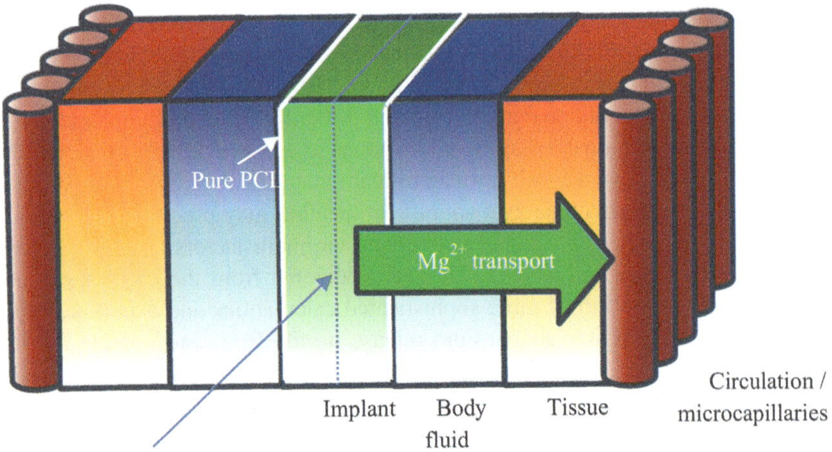

Fig. 1. Geometric arrangement of the model: Implant is a large sheet of PCL containing $MgSO_4{\cdot}7H_2O$ that is optionally coated with a layer of pure PCL. The implant is in contact with body fluid followed by a layer of tissue. Finally, microcapillaries are present that transport the excess magnesium away from the implant area. The transport is one dimensional; the setup is symmetric, and calculation is done only for half of the system.

presentation [5]. Briefly, PCL (M = 80,000 g/mol) was loaded with approximately 20w% $MgSO_4{\cdot}7H_2O$ by cryo-milling the polymer pellets with sieved salt crystals to exclude sizes above 45 μm. The resulting powder was then extruded to form filaments with varied radii. Additionally, flat samples of varied thickness were made from filaments by pressing the material between two flat surfaces lined with PTFE sheets above the melting point of PCL. The release of Mg^{2+} into water was then measured at 37 °C using electrical conductivity. The resulting release profile was fitted to theoretical diffusion driven release profiles [6] using least squares optimization. However, it was found that the release of $MgSO_4$ did not follow a simple diffusion process, and it was necessary to assume two parallel processes with two distinct diffusion coefficients. Thus, Eq. 1 was used to fit the magnesium release from filament samples and Eq. 2 for sheet samples by combining two parallel diffusion processes.

$$F_c(t) = 1 - f_1 \sum_{n=1}^{\infty} \frac{4}{R^2\alpha_n^2} e^{-D_1\alpha_n^2 t} - f_2 \sum_{n=1}^{\infty} \frac{4}{R^2\alpha_n^2} e^{-D_2\alpha_n^2 t} \tag{1}$$

$$F_s(t) = 1 - f_1 \sum_{n=0}^{\infty} \frac{8}{(2n+1)^2\pi^2} e^{-\frac{D_1(2n+1)^2\pi^2 t}{4L^2}} - f_2 \sum_{n=0}^{\infty} \frac{8}{(2n+1)^2\pi^2} e^{-\frac{D_2(2n+1)^2\pi^2 t}{4L^2}} \tag{2}$$

where F_c is the release profile over time (t) for filament (cylinder) and F_s for sheet, expressed as cumulative Mg^{2+} fraction released. R is the radius of the filament, α_n is defined through $J_0(R\alpha_n) = 0$, where J_0 is the zeroth order Bessel function of the first kind. D_1 and D_2 are the two diffusion coefficients, and f_1 and f_2 are the corresponding fraction of magnesium released through processes associated with D_1 and D_2, respectively. L is the half-thickness of the sheet. The fitting was carried out by truncating the infinite sums to a number of terms such that use of additional terms gave negligible contribution to the sums.

It was found that the recovered fitting parameters D_1 and D_2 as well as f_1 and f_2 were reasonably consistent across different sample shapes and sample radius / thickness. For filament samples (4 samples): $D_{1,c} = (4.8 \pm 2.2)$ x10^{-9} cm^2/s, $D_{2,c} = (7.8 \pm 2.1)$ x10^{-10} cm^2/s, $f_{1,c} = 0.77 \pm 0.12$, and $f_{2,c} = 0.23 \pm 0.12$. For sheet samples (4 samples): $D_{1,s} = (4.8 \pm 0.4)$ x10^{-9} cm^2/s, $D_{2,s} = (5.5 \pm 3.3)$ x10^{-10} cm^2/s, $f_{1,s} = 0.85 \pm 0.05$, and $f_{2,s} = 0.15 \pm 0.05$. Averaging over eight samples including both geometries and varied thickness/radius gave: $D_{1,avg} = (4.8 \pm 1.4) \times 10^{-9}$ cm^2/s, $D_{2,avg} = (6.7 \pm 2.8) \times 10^{-10}$ cm^2/s, $f_{1,avg} = 0.81 \pm 0.10$, and $f_{2,avg} = 0.19 \pm 0.10$.

To simplify the present simulation of Mg^{2+} release, only a single diffusion process was considered with diffusion coefficient rounded to $D = 5 \times 10^{-9}$ cm^2/s, see Table 1. This is justifiable since approximately 80% of the magnesium sulphate is released through the faster diffusion process which is primarily responsible for the initial magnesium release that corresponds to the concentration elevation to the therapeutic range.

The diffusion coefficient of magnesium sulphate in body fluid was approximated by the value measured in water (-8×10^{-6} cm^2/s at 25 °C) [7]. This value is expected to be lowered somewhat by the elevated ionic strength in body fluid, while the higher temperature (37 °C) would increase the value [8]. Since both effects result in moderate and opposing changes, they were ignored.

Reports on the diffusion coefficient of magnesium ions in tissue are rather rare. However, it was reported that magnesium has $(1.88 \pm 0.09) \times 10^{-6}$ cm^2/s intracellular diffusion coefficient in rat muscle fibers [9], and (2 to 3) $\times 10^{-6}$ cm^2/s in barnacle muscle [10]. A similar value, 2×10^{-6} cm^2/s, was found in a very different tissue type of squid axon [11]. Based on the above reports, a diffusion coefficient of 2×10^{-6} cm^2/s was used for magnesium transport through the tissue layer.

Table 1. Parameters used in the numerical simulation.

Layer	$D(Mg^{2+})$, cm^2/s	Initial Mg^{2+} concentration, mM	Layer thickness values used, μm
Implant	5×10^{-9}	480, 990	300, 600
PCL coating	5×10^{-9}	0	0, 10
Body fluid	8×10^{-6}	1	0, 10, 20, 50
Tissue	2×10^{-6}	1	20, 50, 80
Circulation	Not Applicable	1	Not Applicable

2.4 Numerical Simulation Method

A very simple method was used for the simulation as described by Crank in his book [6] in Chapter 8, Sect. 8.3. This consisted of calculating the change in concentrations for a small time step, τ, then updating the concentrations with the changes. To avoid divergence, the time step was selected such that $\tau < 0.5h^2/D$ where h is the layer thickness used as the distance step ($h = 1\ \mu m$), and D is the diffusion coefficient in that layer. The calculation was carried out using Python code, and total time span was 87000 s.

3 Results

3.1 Influence of Layer Thickness Variations

It was found that the thicknesses of implant material, PCL coating on the implant, and tissue layer have significant influence on the evolution of magnesium concentrations within the tissue. On the other hand, the thickness of the body fluid layer, having the fastest magnesium diffusion, had small influence on the results. Thus, most of the results are presented for the case when the body fluid layer is the thickest (50 μm).

3.2 Evolution of Magnesium Concentration in the Implant

The change in magnesium concentration within the implant is shown in Fig. 2 for 480 mM initial Mg loading. The curves have the expected shapes based on diffusion theory [6]. Addition of 10 μm thick pure PCL layer has negligible effect on the overall evolution of concentrations except for limiting the maximum magnesium concentration on the surface in contact with the body fluid. This is also expected, as the presence of a pure PCL layer leads to a short lag time [12] that is sufficient to moderate the surface concentration since it provides time for the surroundings to take up magnesium as it is appearing on the surface from the implant.

Similarly, changing the layer thickness of body fluid and tissue in contact has negligible influence on the shape of the curves within the implant (not shown in the figure). Calculations using 990 mM initial concentration provided the same shape of curves as shown in Fig. 2, except the concentration values were higher, proportional to the initial magnesium concentration.

The effect of using a thicker PCL implant is significant. The amount of magnesium remaining in the implant after a given duration is much higher than in the case of thinner implant. However, the magnesium concentrations on the implant surface are comparable for thin and thick implants. Again, the results are in full agreement with theoretical equations [6].

3.3 Concentration Profiles in Body Fluid and Tissue

The calculated concentration profiles of Mg^{2+} over time at selected locations are plotted in Fig. 3, Fig. 4, Fig. 5 to Fig. 6 for different conditions. To limit the number of plots, only the cases with 80 μm tissue thickness were included. The presence of thinner tissue layers accelerates the decrease of magnesium concentration over time. The locations plotted are body fluid in contact with the implant, tissue surface facing the implant, tissue middle layer, and tissue facing capillaries. Of these, the tissue surface in contact with the capillaries has insignificant magnesium concentration increase in all cases.

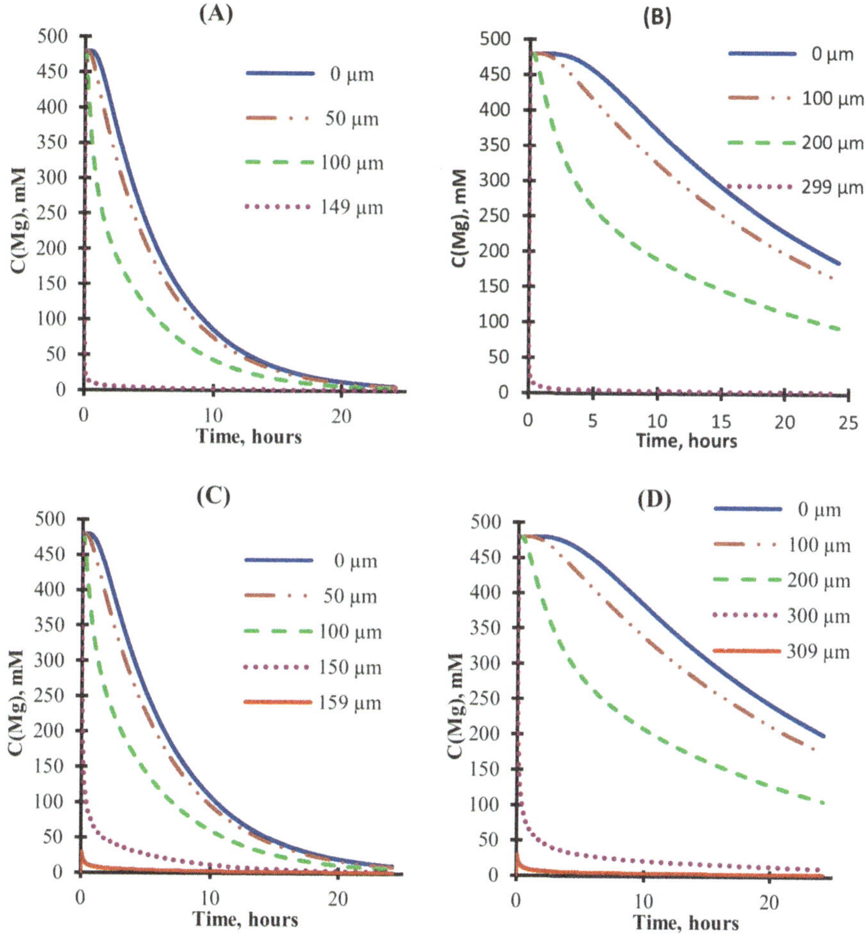

Fig. 2. Evolution of magnesium concentration within implant at different locations (center at 0 μm, surface at 149 μm or 299 μm) for 480 mM initial Mg^{2+} loading, 50 μm body fluid layer and 80 μm tissue layer. (A): No PCL coating is present, 300 μm thick implant. (B): No PCL coating is present, 600 μm thick implant. (C): 10 μm PCL coating without magnesium is present on 300 μm thick implant; concentration on coating layer surface rises fast to ~30 mM, then decays (lowest curve in plot). (D): 10 μm PCL coating without magnesium is present on 600 μm thick implant; concentration on coating layer surface rises fast to ~30 mM, then decays (lowest curve in plot).

The time dependent concentration profiles indicate a very sharp rise in magnesium level within less than a minute if no pure PCL coating is present on the implant. If 10 μm

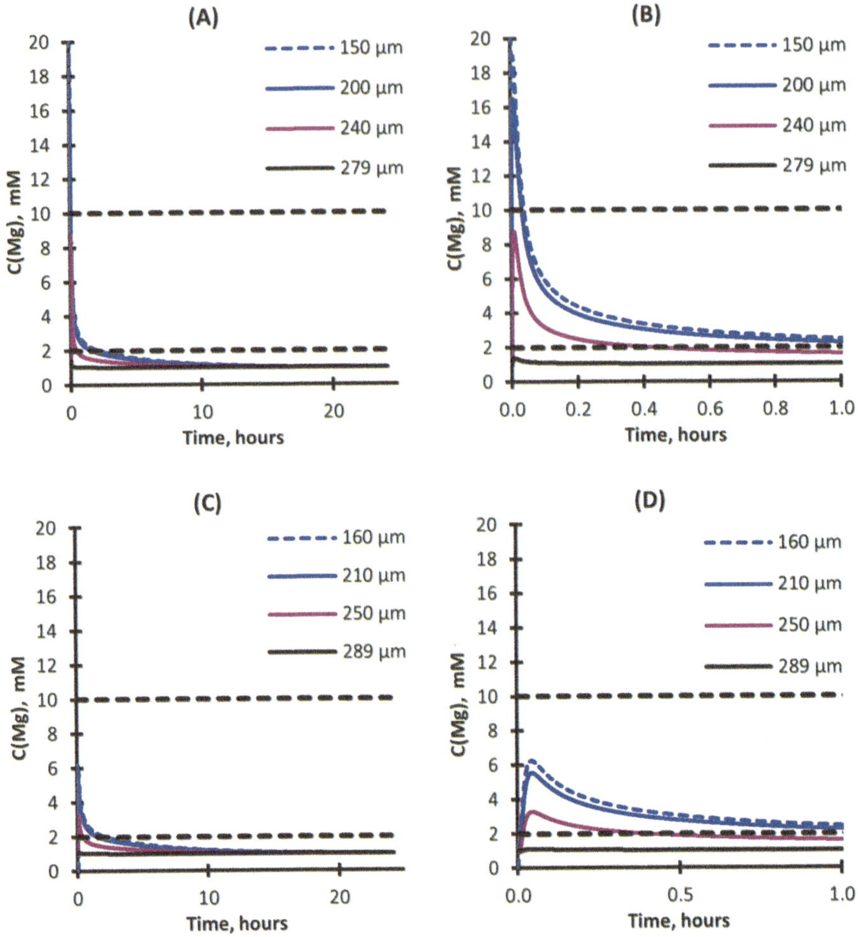

Fig. 3. Evolution of Mg^{2+} concentration in body fluid at the implant surface (dashed blue lines) and in tissue (solid lines indicating different locations) with the lower 480 mM initial Mg^{2+} loading in the implant. Implant thickness is 300 μm. The optimum therapeutic concentration range is indicated by the black horizontal dashed lines. (A) and (B) with different time scales: No pure PCL coating on the implant showing highly elevated magnesium level in the initial few minutes. (C) and (D) with different time scales: 10 μm pure PCL coating is present on the implant, significantly moderating the initial magnesium concentration rise, avoiding buildup of potentially toxic levels.

PCL coating is added, the peak magnesium concentration is reached within about three minutes and this peak level is much lower than in the case where no PCL coating is in place. After the peak concentration is reached, the concentration drops quickly with time evolution resembling a decaying exponential function.

The effect of initial magnesium loading in the implant is the shifting of concentration values in proportion to the initial concentration without otherwise affecting the profiles (compare Fig. 3 and Fig. 4), as seen similarly in the case of concentrations within the implant.

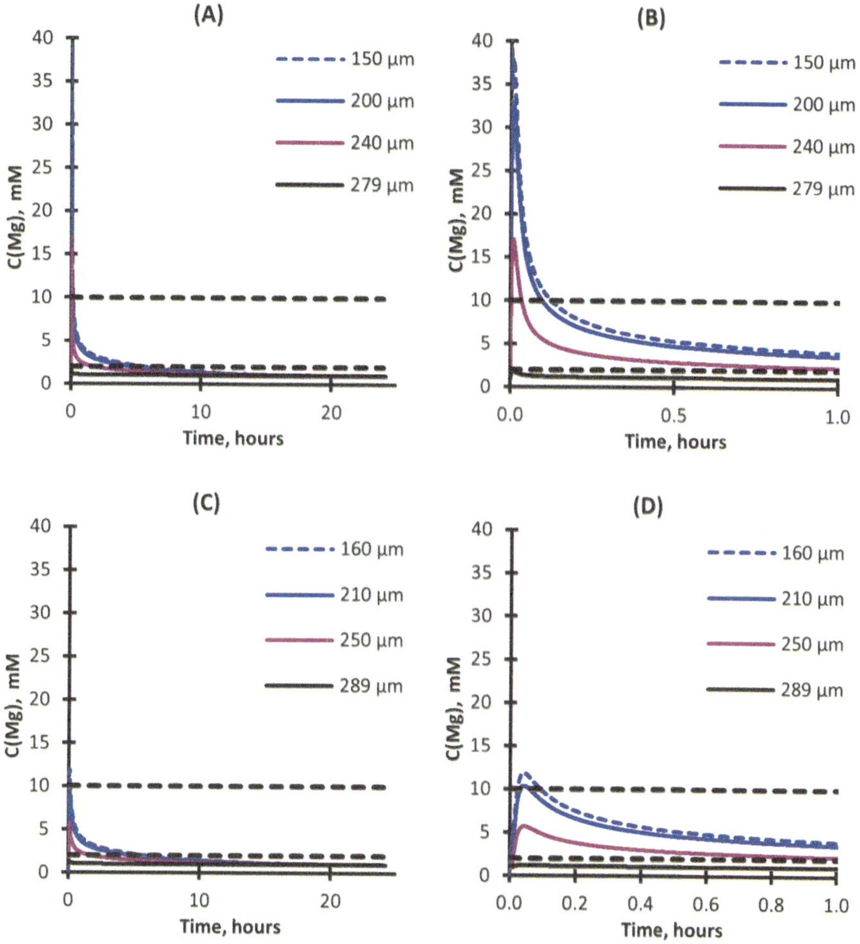

Fig. 4. Evolution of Mg^{2+} concentration in body fluid at the implant surface (dashed blue lines) and in tissue (solid lines indicating different locations) with higher 990 mM initial Mg^{2+} loading in the implant. Implant thickness is 300 μm. The optimum therapeutic concentration range is indicated by the black horizontal dashed lines. (A) and (B) with different time scales: No pure PCL coating on implant showing highly elevated magnesium level in the initial few minutes. (C) and (D) with different time scales: 10 μm pure PCL coating is present on the implant which significantly moderates the initial magnesium concentration rise, thus avoiding buildup of potentially toxic levels. The shapes of the curves are nearly identical to those in Fig. 3 except for the concentration values are higher in proportion to the initial loading level in the implant.

The thickness of the implant has no effect on the initial part of the concentration profiles since the initial magnesium release is determined by the magnesium concentration near the surface of the implant. However, as the magnesium content is depleted in the

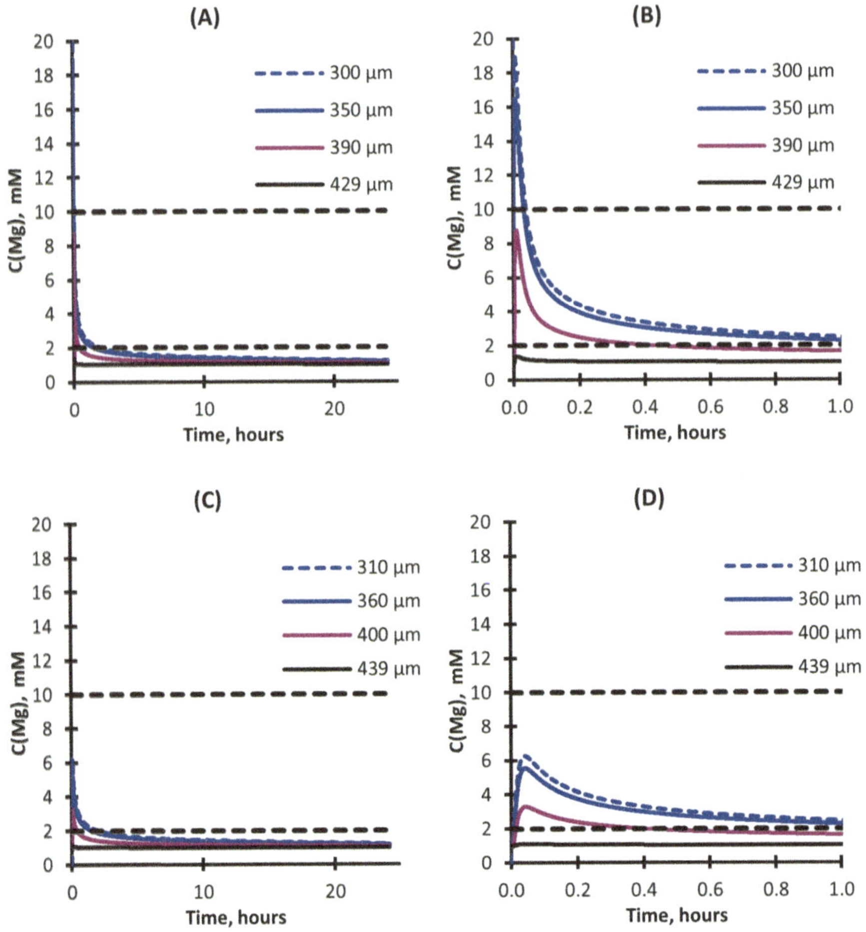

Fig. 5. Evolution of Mg^{2+} concentration in body fluid at the implant surface (dashed blue lines) and in tissue (solid lines indicating different locations) for thicker implant of 600 μm, and low salt content of 480 mM initial Mg^{2+} loading in the implant. The optimum therapeutic concentration range is indicated by the black horizontal dashed lines. (A) and (B) with different time scales: No pure PCL coating on implant showing highly elevated magnesium level in the initial few minutes. (C) and (D) with different time scales: 10 μm pure PCL coating is present on the implant which significantly moderates the initial magnesium concentration rise, thus avoiding buildup of potentially toxic levels.

implant over time, the thinner material starts to supply a lesser amount of magnesium while the thicker implant can supply more magnesium as it diffuses from the interior. Thus, thicker implant increases the magnesium concentration in the tissue layer at later times of the release.

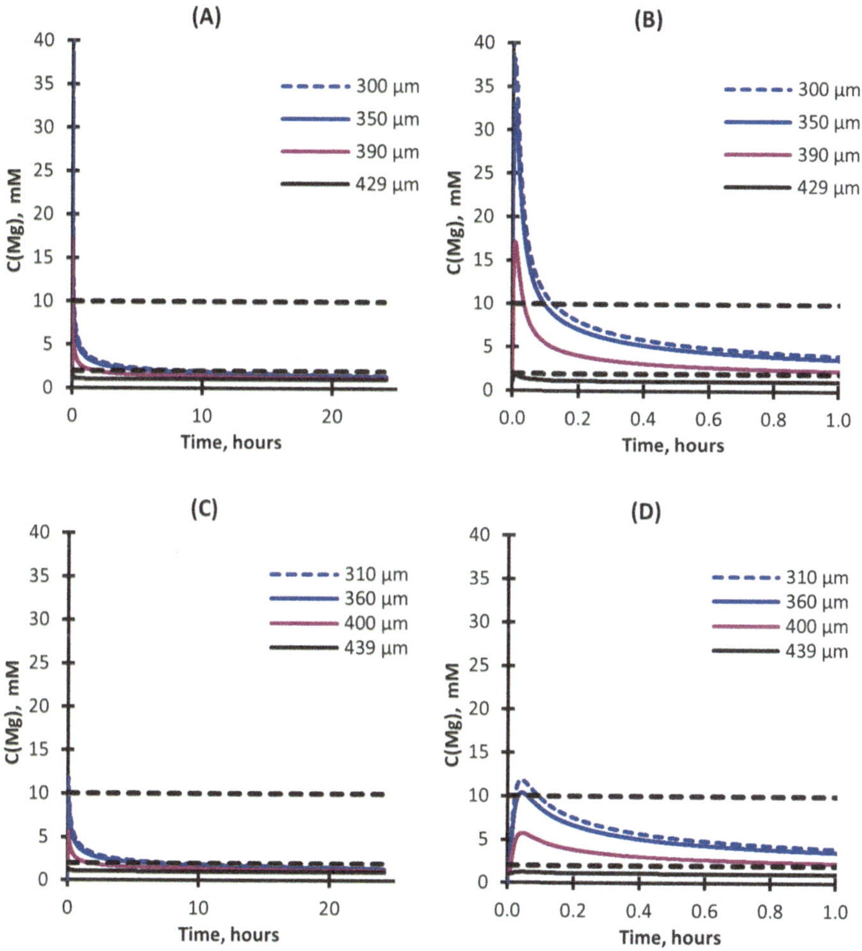

Fig. 6. Evolution of Mg^{2+} concentration in body fluid at the implant surface (dashed blue lines) and in tissue (solid lines indicating different locations) for thicker implant of 600 μm, and higher salt content of 990 mM initial Mg^{2+} loading in the implant. The optimum therapeutic concentration range is indicated by the black horizontal dashed lines. (A) and (B) with different time scales: No pure PCL coating on implant showing highly elevated magnesium level in the initial few minutes. (C) and (D) with different time scales: 10 μm pure PCL coating is present on the implant which significantly moderates the initial magnesium concentration rise, thus avoiding buildup of potentially toxic levels. The shapes of the curves are nearly identical to those in Fig. 5 except for the concentration values are higher in proportion to the initial loading level in the implant.

4 Discussion

The key parameters that can be obtained from the simulation are the duration of elevated magnesium concentration within the optimum therapeutic range (2 mM to 10 mM), and the duration of magnesium concentration exceeding the optimum therapeutic concentration which may potentially lead to toxic levels of Mg^{2+}. Since the thickness of the body fluid layer (0, 10, 20, and 50 μm) has little effect on the duration of elevated concentrations in the tissue, the duration values were averaged for the cases with varied body fluid layer thickness. The data is plotted in Figs. 7 and 8 to indicate the main factors responsible for shortening/extending the magnesium concentration in the potentially toxic range and in the therapeutic range.

As seen in Fig. 7, the duration of therapeutic magnesium concentration is rather short, and it is strongly influenced by the thickness of tissue between the implant and capillaries. The initial magnesium concentration in the implant is also a major factor in controlling the duration. The higher the initial magnesium concentration in the implant, the longer the duration of elevated concentration. The implant thickness has moderate effect, and it depends on the tissue thickness and the initial magnesium loading, while the presence of a thin pure PCL coating layer has negligible effect on the duration of therapeutic concentration.

The duration of elevated magnesium concentration is also strongly dependent on the location within the tissue. Several fold shorter durations are seen in the center of the tissue layer as compared the surface facing the implant.

The magnesium concentration exceeding the optimum level is restricted to a few minutes only as seen in Fig. 8. The excessively high concentration is extended if thicker tissue layer is present or when higher initial magnesium loading is applied in the implant. A very significant effect is seen due to the application of a pure PCL layer on the implant which can reduce or eliminate the duration when magnesium concentration is too high. Implant thickness has no effect on the duration of excess magnesium level in the investigated range.

Although, the simulation provided clear insights into the temporal and spatial evolution of magnesium concentration in the vicinity of a magnesium releasing implant, it must be highlighted that there are several shortcomings of the simple model used. Some of these limitations include the assumption of homogeneous tissue layer instead of a cellular structure, the uncertainty of the diffusion coefficients used, and assumption of a purely diffusional magnesium transport between implant and capillaries when there is also possibility of active transport of magnesium into and out of cells, as well as storage of some of the magnesium in the cells. It can also be considered that tissue layers are not solid slabs, but regions separated by small channels where body fluid is present. In such a case, the body fluid may be a fast conduit for magnesium diffusion, thus reducing the overall concentration increase.

In light of the limitations, the results should be taken as semiquantitative that can guide experimental designs on the measurement efficacy of implants or provide a basis for more sophisticated numerical analysis.

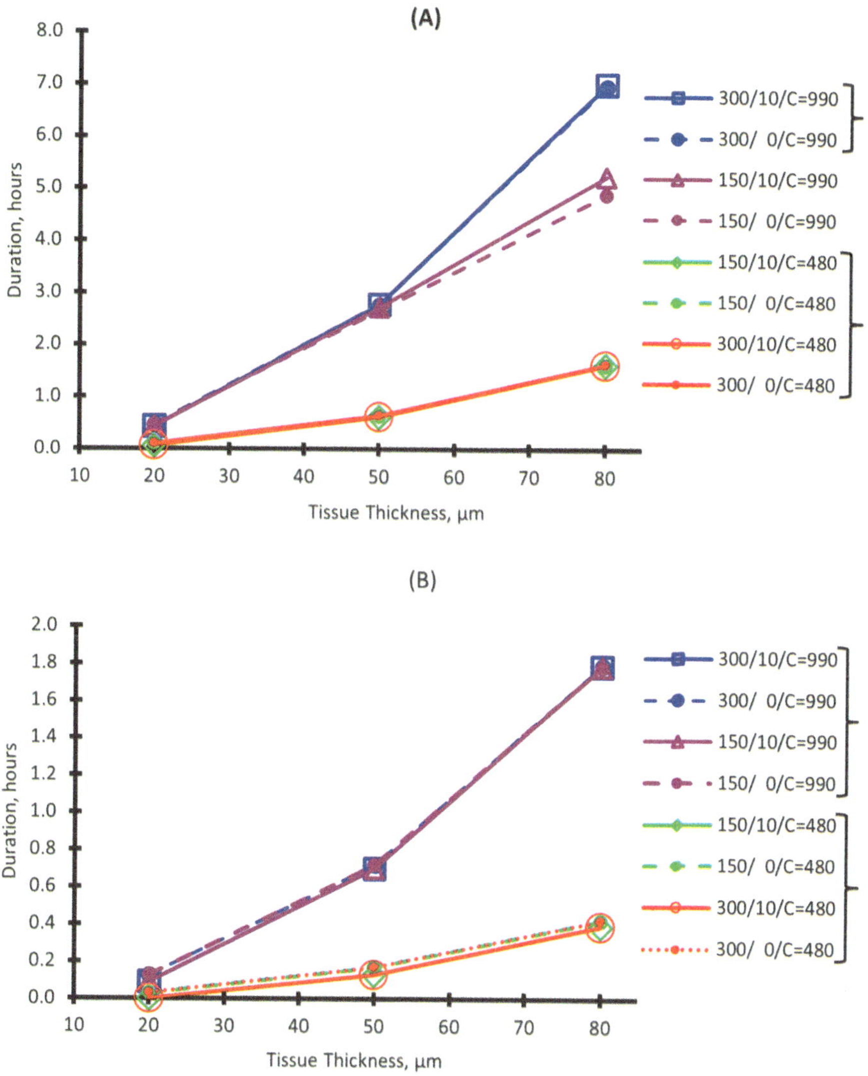

Fig. 7. Duration of Mg^{2+} concentration within the therapeutic range (averaged over varied body fluid layer thickness). Legend: xxx/yy/C = zzz where xxx is the half thickness of the implant, yy is the thickness of pure PCL coating layer, and zzz is the initial Mg^{2+} concentration in the implant in mM; curly brackets indicate data that nearly or fully coincide. (A): Therapeutic Mg^{2+} concentration at the surface of the tissue facing the implant. (B): Therapeutic Mg^{2+} concentration at the center of the tissue layer.

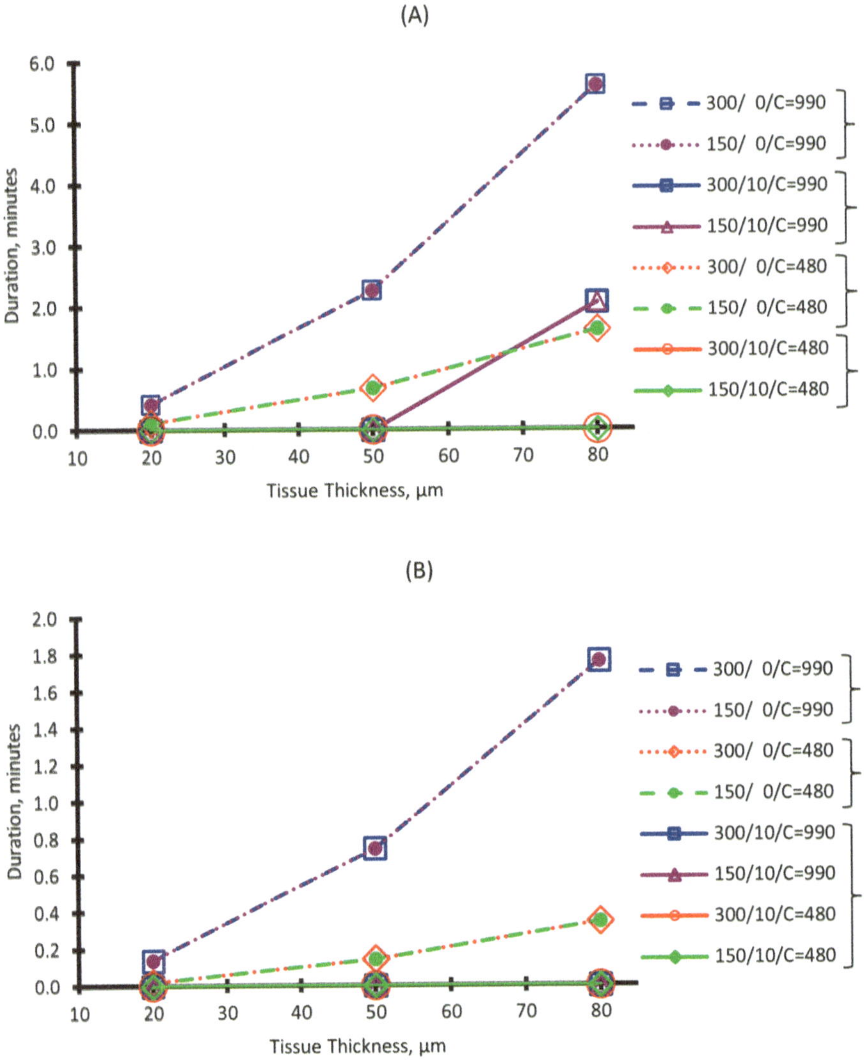

Fig. 8. Duration of Mg^{2+} concentration exceeding the optimum therapeutic range (averaged over varied body fluid layer thickness). Legend: xxx/yy/C = zzz where xxx is the half thickness of the implant, yy is the thickness of pure PCL coating layer, and zzz is the initial Mg^{2+} concentration in the implant in mM; curly brackets indicate data that nearly or fully coincide. (A): Excessive Mg^{2+} concentration at the surface of the tissue facing the implant. (B): Excessive Mg^{2+} concentration at the center of the tissue layer.

5 Conclusions

A simple model was set up to numerically evaluate the diffusion controlled release of magnesium sulphate from polycaprolactone implant. The results indicated that the magnesium is released from the implant at a moderate rate, then it is transported by diffusion across a body fluid layer between implant and tissue at a high rate since the magnesium diffusion coefficient is highest in the body fluid. The subsequent tissue layer with a moderate diffusion coefficient then poses a slight barrier and slows down the diffusion of magnesium.

Two key parameters were identified to describe the efficacy of magnesium release from the implant: The duration of magnesium concentration in the tissue when the magnesium is in the optimum therapeutic range, and the duration where magnesium concentration is above the optimum therapeutic range.

It was found that the duration of optimum magnesium concentration was rather short, well below one day in the scenarios investigated. It can be extended by increasing the initial magnesium loading in the implant, and by using a thicker implant that exhausts its magnesium supply over a longer period. Additionally, the presence of thicker tissue layers between implant and capillaries increased the duration of elevated magnesium concentration in the tissue. On the other hand, the presence of pure polycaprolactone coating had negligible effect on the duration of optimum concentration.

Duration of very high and potentially toxic magnesium levels was found to be short, less than six minutes in all investigated scenarios. This duration is shorter with thinner tissue layers, and for lower initial magnesium loading in the implant. The implant thickness had no effect on this duration but the presence of a pure polycaprolactone coating on the implant had a major effect. Such a coating limits the maximum surface concentration of magnesium in the implant and introduces the magnesium into the body fluid / tissue slower in the initial few minutes.

Thus, the simulation indicated that designing an implant with extended elevated magnesium concentration would rely on high initial magnesium concentration in the implant (>1 M), preferably with thick implants (>600 μm). Additionally, a thin layer of pure polycaprolactone layer (10 to 20 μm) can be applied on the implant to mitigate or prevent the initial magnesium concentration surge when the implant first comes into contact with the tissue.

Overall, it was found that a highly soluble salt such as magnesium sulphate can be incorporated into polycaprolactone implant to achieve elevated magnesium concentration for a limited time period that can promote bone regeneration. Due to the fast clearing of magnesium from the tissue, it was seen that toxic levels of magnesium may be present only for a short time of a few minutes which can be substantially reduced or eliminated by applying a pure PCL layer on the implant.

Acknowledgements. This research is supported by A*STAR under its RIE2025 Manufacturing, Trade and Connectivity (MTC) Industry Alignment Fund Industry Collaboration Projects (IAF-ICP) (Award I2101E0010). The support of Osteopore International Pte. Ltd. is also acknowledged.

References

1. De Baaij, J.H.F., Hoenderop, J.G.J., Bindels, R.J.M.: Magnesium in man: implications for health and disease. Physiol. Rev. **95**, 1–46 (2015). https://doi.org/10.1152/physrev.00012.2014

2. Yuan, Z., Wan, Z., Gao, C., Wang, Y., Huang, J., Cai, Q.: Controlled magnesium ion delivery system for in situ bone tissue engineering. J. Controlled Release **350**, 360–376 (2022)
3. Tsai, P.S., et al.: Correlations of neuronal and microvascular densities in murine cortex revealed by direct counting and colocalization of nuclei and vessels. J. Neurosci. **29**(46), 14553–14570 (2009)
4. Arthur, E., Elsner, A.E., Sapoznik, K.A., Papay, J.A., Muller, M.S., Burns, S.A.: Distances from capillaries to arterioles or venules measured using OCTA and AOSLO. Invest. Ophthalmol. Vis. Sci. **60**, 1833–1844 (2019)
5. Nemeth, S., Loke, C.: Drug release times from different shapes of releasing bodies. In: The 18th International Conference on Biomedical Engineering, 9–12 December 2024, Singapore, Contribution Number: 199 (2024)
6. Crank, J.: The Mathematics of Diffusion, 2nd edn. Clarendon Press, Oxford (1975)
7. Harned, H.S., Hudson, R.M.: The diffusion coefficient of magnesium sulfate in dilute aqueous solution at 25°. J. Am. Chem. Soc. **73**(12), 5880–5882 (1951)
8. Onsager, L., Fuoss, R.M.: Irreversible processes in electrolytes, diffusion, conductance, and viscous flow in arbitrary mixtures of strong electrolytes. J. Phys. Chem. **36**(11), 2689–2848 (1932)
9. Bernengo, J.C., Collet, C., Jacquemond, V.: Intracellular Mg^{2+} diffusion within isolated rat skeletal muscle fibers. Biophys. Chem. **89**, 35–51 (2001)
10. Ashley, C.C., Ellory, J.C.: The efflux of magnesium from single crustacean muscle fibers. J. Physiol. **226**, 653–674 (1972)
11. Baker, P.F., Crawford, A.C.: Mobility and transport of magnesium in squid giant axons. J. Physiol. **227**, 855–874 (1972)
12. Kalosakas, G., Panagopoulou, P.: Lag Time in diffusion-controlled release formulations containing a drug-free outer layer. Processes **10**, 2592 (2022)

Alteration of Biomechanical Properties in Early-Stage Osteoarthritis: A Study on Elastic Modulus and Permeability of Articular Cartilage

Syaidatul Syakirah Salehuddin[1], Mohd Juzaila Abd Latif[1,2](✉), Mohamad Shukri Zakaria[1,2], Muhamad Noor Harun[3], and Munirah Sha'ban[4]

[1] Fakulti Teknologi dan Kejuruteraan Mekanikal, Universiti Teknikal Malaysia Melaka (UTeM), Hang Tuah Jaya, 76100 Durian Tunggal, Melaka, Malaysia
juzaila@utem.edu.my

[2] Centre for Advanced Research On Energy (CARe), Universiti Teknikal Malaysia Melaka (UTeM), Hang Tuah Jaya, 76100 Durian Tunggal, Melaka, Malaysia

[3] Faculty of Mechanical Engineering, Universiti Teknologi Malaysia (UTM), 81310 Skudai, Johor, Malaysia

[4] Kulliyah of Allied Health Sciences, International Islamic University Malaysia, Bandar Indera Mahkota, 25200 Kuantan, Pahang, Malaysia

Abstract. Osteoarthritis is a degenerative joint disease that progressively impairs the biomechanical functionality of articular cartilage. Although it is typically diagnosed during its advanced stages, subtle biomechanical changes emerge much earlier, preceding any observable morphological damage. Therefore, this study investigates these early-stage biomechanical changes. Cartilage samples were extracted from the bovine femoral head and then subjected to 0.05% trypsin treatment to mimic progressive degradation. The elastic modulus and permeability were determined by integrating the indentation experimental data and finite element analysis. The results demonstrated a marked decline in the elastic modulus, decreasing by 78% from 1.89 MPa to 0.42 MPa. Conversely, permeability significantly increased by 201%, rising from 1.34×10^{-15} m^4/Ns to 4.03×10^{-15} m^4/Ns. The outcomes enhance understanding of the biomechanical changes associated with early osteoarthritis and offer valuable insights for developing quantitative diagnostic tools for early-stage detection.

Keywords: Osteoarthritis · Articular Cartilage · Elastic Modulus · Permeability

1 Introduction

Osteoarthritis is a degenerative joint disorder characterized by the deterioration of articular cartilage within synovial joints. This disease affects over 30% of individuals aged 45 years and older and is a leading cause of chronic disability worldwide [1]. The condition significantly impairs mobility and negatively impacts daily activities [2]. Early detection of osteoarthritis is crucial for implementing effective treatments, as the disease is often identified only in its advanced stages [3].

C. T. Lim et al. (Eds.): ICBME 2024, IFMBE Proceedings 132, pp. 69–78, 2025.
https://doi.org/10.1007/978-3-032-04372-6_6

The changes in the pathology of osteoarthritis start from the degradation of articular cartilage, subchondral damage, and eventually bone remodeling [4]. These can also be classified into four stages of osteoarthritis, starting from early, mild, moderate, and severe. During the early stage of osteoarthritis, the components in the cartilage gradually degrade which causes the increment of water content [5]. Previous studies have found that the decrease in biomechanical properties occurs before noticeable morphological changes in cartilage during the early stages of osteoarthritis, without any significant alterations in thickness [6, 7]. It is unclear whether the decrease in biomechanical properties precedes or follows structural changes in cartilage degeneration, as the relationship between them needs to be better defined [8]. Furthermore, swelling of the cartilage tissue has been identified as an early feature of osteoarthritis, occurring prior to its thinning and degradation [3].

While osteoarthritis progresses through various stages, much of the research on articular cartilage has focused on later-stage disease. Methods commonly employed to induce osteoarthritis in experimental studies include mechanical, surgical, and chemical degradation [9]. However, these approaches are not strictly associated with specific disease stages, resulting in limited research on the early stages of osteoarthritis. This study seeks to address this gap by examining the biomechanical behavior of the cartilage tissue during the onset of osteoarthritis progression.

2 Materials and Method

2.1 Sample Preparation

Intact bovine hip joints ($n = 6$) were procured from a local abattoir within 12 h post-slaughter. The femoral heads were meticulously separated and the extraneous tissues were removed. Cylindrical cartilage specimens ($n = 60$) were extracted from the femoral heads using an 8-mm diameter hollow drill bit, as shown in Fig. 1. The samples were then allocated into two groups: control ($n = 12$) and degraded ($n = 48$). During preparation, the samples were consistently moistened using phosphate-buffered saline (PBS) rinses to prevent dehydration. All specimens were maintained at 10 °C, wrapped in PBS-moistened tissues, to ensure their integrity prior to analysis.

2.2 Preparation of Early-Stage Osteoarthritic Cartilage Specimens

Cartilage specimens ($n = 60$) were equilibrated by immersing test tubes in a water bath at 37 °C for 60 min. During the last 30 min of equilibration, test tubes containing 0.05% trypsin solution were also immersed in the water bath. Trypsin was used to enzymatically degrade proteoglycans in cartilage, targeting the hallmark biochemical changes associated with early osteoarthritis while preserving tissue morphology [10]. Since proteoglycan degradation is a hallmark of early-stage osteoarthritis, trypsin was deemed the optimal enzyme for this study. Cartilage specimens ($n = 48$) were exposed to trypsin in increments of 30 min, up to a maximum of 120 min as shown in Fig. 2. Following each degradation interval, the samples were rinsed in PBS for 5 min and submerged in PBS for 1 h before undergoing creep indentation testing.

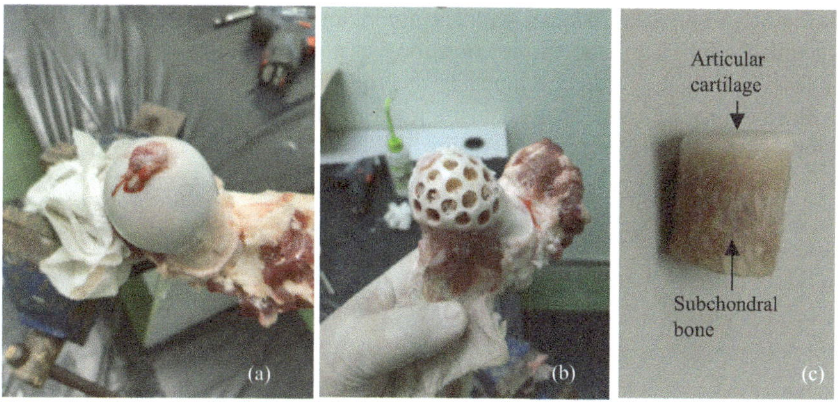

Fig. 1. (a) Bovine femoral head, (b) drilled femoral head and (c) cartilage sample.

Fig. 2. Degradation process in water bath.

2.3 Measurement of Water Content

Cartilage samples ($n = 30$) were separated from the subchondral bone to record the initial wet weight. At every degradation interval, the samples were immersed in PBS for 5 min followed by immersion in normal saline for 30 min. The wet weight of the sample was then measured using an analytical balance scale. The degraded samples were subsequently air-dried at room temperature for 12 h, and their dry weights were recorded. The water content was determined based on the percentage difference between the initial wet weight and dry weight, divided by the initial wet weight [11, 12].

2.4 Characterization of Biomechanical Properties

A creep indentation test was conducted on individual cartilage specimens using the setup shown in Fig. 3. During the experiment, the samples were submerged in PBS to prevent dehydration. A spherical indenter with a 4-mm diameter and 0.18 N compressive load was applied on the cartilage surface, resulting in a strain of 10–20% of the cartilage thickness. Deformation was continuously recorded at 0.01-s intervals for over 2500 s to achieve the equilibrium. Data acquisition was performed using LabVIEW software (National Instruments Corporation, Austin, TX, USA).

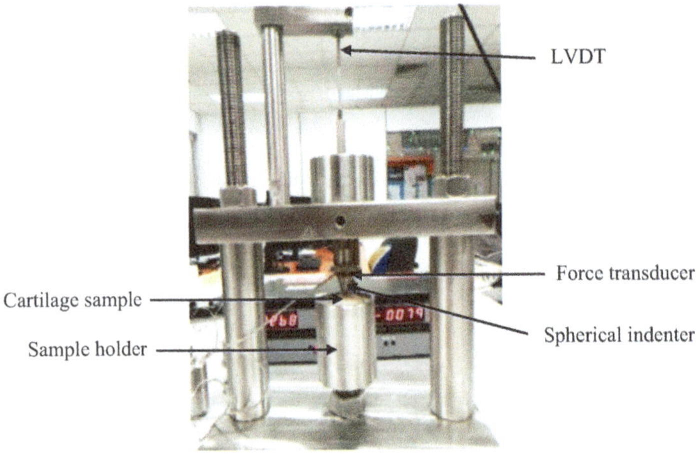

Fig. 3. Cartilage experimental setup.

Articular cartilage was modeled as an axisymmetric biphasic poroelastic material to simulate its solid and fluid phases, utilizing Abaqus software (DS Simulia Corp., Providence, RI, USA). The finite element (FE) model incorporated the measured cartilage thickness of each sample and the dimensions of the subchondral bone, which were 4 mm in width and 2 mm in height, as illustrated in Fig. 4. The spherical indenter was defined as an analytical rigid surface with a 2-mm radius. For the cartilage, four-node bilinear displacement and pore pressure elements (CAX4P) were used, while four-node bilinear elements (CAX4) were employed for the subchondral bone with an elastic modulus of 2000 MPa and a Poisson's ratio of 0.2 [13].

Boundary and interface conditions were defined to replicate the experimental creep indentation test. Horizontal movement of nodes along the axis of symmetry was constrained, while the bottom nodes of the subchondral bone were fixed in both horizontal and vertical directions. Only vertical displacement was permitted for the indenter. Fluid flow constraints were applied, allowing contact-dependent flow at the cartilage surface, while the fluid movement was confined at the base and along the symmetry axis [13]. Zero pore pressure was applied at the cartilage's outer edge to enable unrestricted fluid flow.

The biphasic biomechanical properties, including elastic modulus and permeability, were determined by fitting the deformation-time data obtained from the indentation test to the finite element model. These properties were adjusted iteratively within the model until the simulated deformation-time curve matched the experimental data. Curve

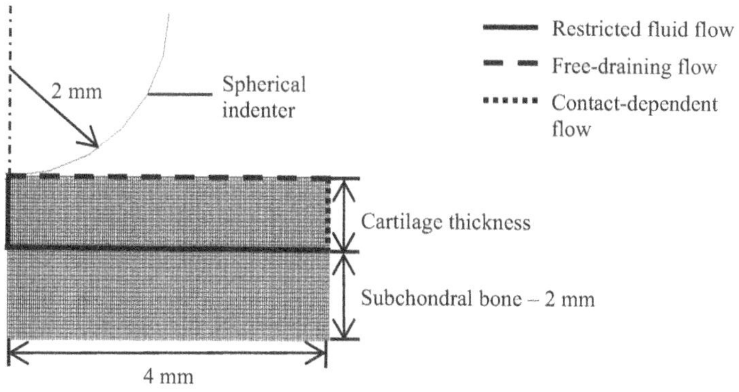

Fig. 4. FE model of cartilage indentation test.

fitting was performed using a non-linear least-squares optimization approach in Matlab software (MathWorks Inc., Natick, MA, USA) [13].

3 Results and Discussion

3.1 Water Content of Cartilage

The initial water content of the control cartilage samples was measured at 73% ± 3% (mean ± standard deviation), which was consistent with previously reported data [7, 14]. As shown in Fig. 5, water content increased with degradation time, rising from 75% ± 5% at 30 min to 88% ± 4% at 120 min. This 21% increase aligns with earlier findings that link water content increases to early-stage osteoarthritis [15]. The increase in water content was linked to the depletion of proteoglycans, which are essential for maintaining the cartilage's water-retention capacity [10]. Proteoglycans retain water by binding water molecules to their negatively charged glycosaminoglycan (GAG) chains. Consequently, the loss of proteoglycans disrupts this binding, leading to an increase in unbound water within the cartilage matrix and a subsequent rise in overall water content.

3.2 Indentation Test Deformation

The deformation characteristics of degraded cartilage samples under early osteoarthritic conditions are presented in Fig. 6. The results demonstrate that cartilage deformation increased progressively with the extent of degradation. At equilibrium, the control samples exhibited a mean deformation of 0.11 ± 0.02 mm (mean ± standard deviation). In contrast, the samples showed a 55% increase in deformation to 0.17 ± 0.03 mm after 30 min, and a 155% increase to 0.28 ± 0.06 mm after 120 min of degradation.

Past studies have established that there is an increment in cartilage water content during the early stages of osteoarthritis [5, 16]. This elevated water content accelerates cartilage deformation during creep indentation testing. Under constant loading conditions, water is expelled from the cartilage matrix until equilibrium stress is achieved [17]. With prolonged degradation, trypsin enzymatically breaks down proteoglycan which further

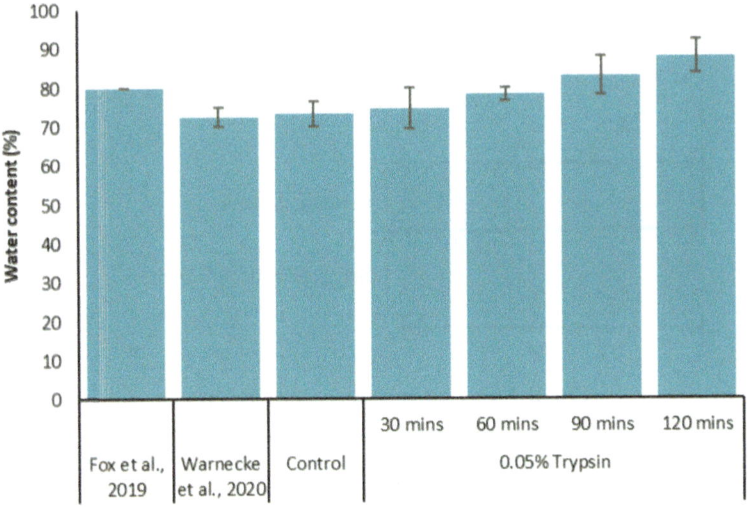

Fig. 5. Water content for control and osteoarthritic samples.

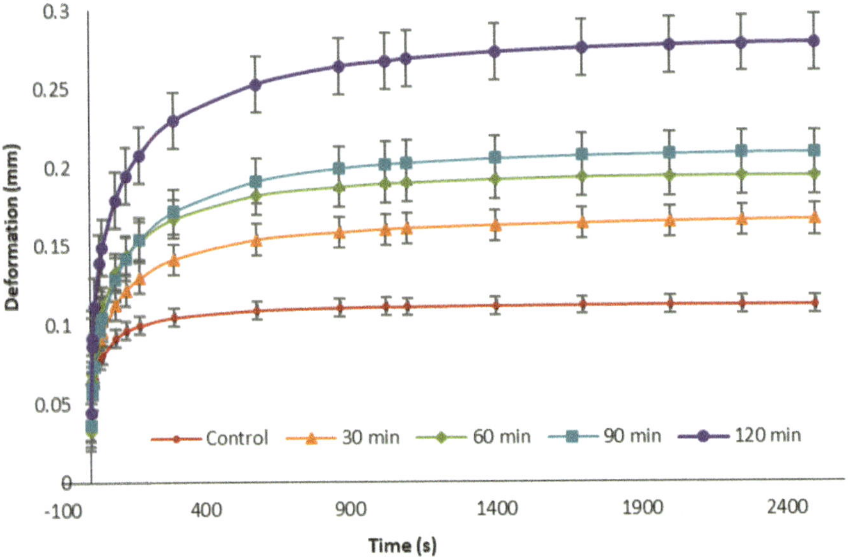

Fig. 6. Creep deformation for control and osteoarthritic samples.

elevates the water content within the cartilage [15]. The increased water content facilitates greater fluid exudation from the collagen framework, leading to higher cartilage deformation.

3.3 Biomechanical Properties of Articular Cartilage

The results for the elastic modulus and permeability are presented in Fig. 7(a) and (b), respectively. The elastic modulus of control samples was measured at 1.89 ± 0.54 MPa (mean ± standard deviation), aligning with previously reported values [18, 19]. Following 30 min of degradation, the elastic modulus declined by 44% to 1.01 ± 0.40 MPa and further declined by 78% to 0.42 ± 0.19 MPa after 120 min of degradation.

In contrast, the permeability of the control samples was found to be 1.34 ± 1.29 $\times 10^{-15}$ m^4/Ns (mean ± standard deviation), aligning with previously reported ranges for bovine cartilage [20]. Permeability exhibited an opposite trend. After 30 min of degradation, the permeability increased by 6% to 1.42 ± 0.92 $\times 10^{-15}$ m^4/Ns and further increased by 201% to 4.03 ± 1.83 $\times 10^{-15}$ m^4/Ns after 120 min of degradation. These findings are consistent with earlier studies reporting a notable increase in cartilage permeability associated with glycosaminoglycan (GAG) depletion [16].

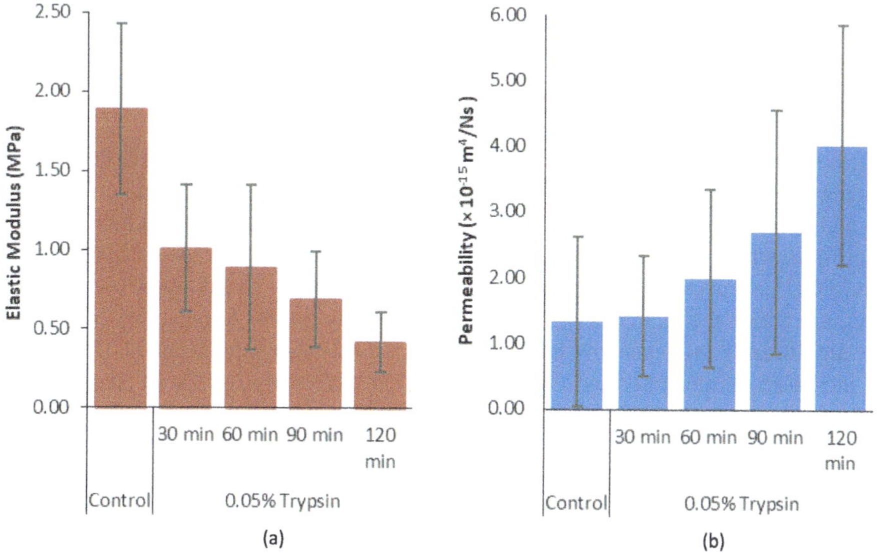

Fig. 7. Biomechanical properties of cartilage at early stages of osteoarthritis: (a) elastic modulus, (b) permeability.

The correlation analysis further supported these relationships, showing a strong positive correlation (r = 0.90) between water content and permeability, and a strong negative correlation (r = − 0.82) between water content and elastic modulus, as shown in Fig. 8(a) and (b).

The GAGs within proteoglycan aggregates play a critical role in determining the pore size of the cartilage matrix, thereby regulating fluid flow resistance. When GAGs are depleted, the pore size increases, leading to reduced fluid flow restriction and consequently elevated cartilage permeability. This phenomenon is linked to an increase in free water content within the extracellular matrix, as proteoglycans are progressively

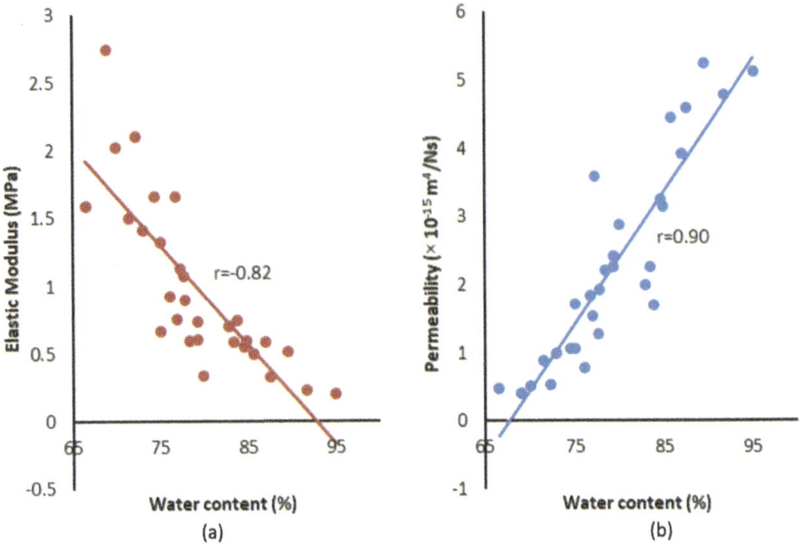

Fig. 8. Correlation between water content and (a) elastic modulus, (b) permeability.

degraded by trypsin. Notably, this increase in water content is observed in the early stages of osteoarthritis before any significant degradation of the collagen network [15]. In addition to their structural role, the negatively charged GAGs generate electrostatic repulsion, which contributes to the fixed charge density (FCD) in the cartilage matrix. The FCD enhances the tissue's compressive resistance by maintaining high osmotic pressure within the matrix [21]. Both the elastic modulus and permeability of cartilage are significantly influenced by the FCD. The loss of GAGs reduces the FCD, resulting in increased permeability and diminished compressive stiffness. These changes compromise the load-bearing capacity of articular cartilage, highlighting the critical role of GAGs in maintaining cartilage biomechanical functionality.

The increase in water content caused by proteoglycan depletion diminishes the cartilage's ability to resist deformation. This loss of structural integrity leads to a reduced elastic modulus, as the tissue can no longer efficiently distribute or resist compressive loads. This was portrayed by the strong negative correlation between water content and elastic modulus ($r = -0.82$) in Fig. 8(a). Conversely, permeability exhibits a strong positive correlation with water content ($r = 0.90$) as shown in Fig. 8(b). Higher water content increases matrix porosity due to the degradation or displacement of structural components, thereby reducing resistance to fluid flow. The resulting increase in permeability alters the cartilage's ability to dynamically distribute mechanical loads, potentially exacerbating degenerative processes under repetitive stress. These findings underscore the critical interplay between matrix composition, fluid dynamics, and mechanical functionality in articular cartilage.

4 Conclusion

In the present study, the quantitative assessment of the biomechanical properties of articular cartilage at early-stage osteoarthritis has provided valuable insights into how this degenerative disease impacts the mechanical behavior of cartilage. The outcome demonstrated significant alterations in biomechanical properties, particularly in terms of permeability, following degradation at 120 min compared to the elastic modulus. These findings offer a deeper understanding of the changes that occur in cartilage during the early stages of osteoarthritis, which could be crucial for developing strategies for early detection and intervention of the disease.

Acknowledgment. The authors gratefully acknowledge the financial support provided by the Ministry of Higher Education Malaysia through the Fundamental Research Grant Scheme (FRGS/1/2021/TK0/UTEM/02/41). Additional support from Universiti Teknikal Malaysia Melaka (UTeM) is also deeply appreciated.

References

1. Collins, A.T., et al.: Selective enzymatic digestion of proteoglycans and collagens alters cartilage T1rho and T2 relaxation times. Ann. Biomed. Eng. **47**, 190–201 (2019). https://doi.org/10.1007/s10439-018-02143-7
2. Winalski, C.S., Rajiah, P.: The evolution of articular cartilage imaging and its impact on clinical practice. Skeletal Radiol. **40**, 1197–1222 (2011). https://doi.org/10.1007/s00256-011-1226-z
3. Hayashi, D., Roemer, F.W., Guermazi, A.: Imaging for osteoarthritis. Ann. Phys. Rehabil. Med. **59**, 161–169 (2016). https://doi.org/10.1016/j.rehab.2015.12.003
4. Fang, T., Zhou, X., Jin, M., Nie, J., Li, Xi, I.: Molecular mechanisms of mechanical load-induced osteoarthritis. Int. Orthop. **45**, 1125–1136 (2021). https://doi.org/10.1007/s00264-021-04938-1
5. Mieloch, A.A., Richter, M., Trzeciak, T., Giersig, M., Rybka, J.D.: Osteoarthritis severely decreases the elasticity and hardness of knee joint cartilage: a nanoindentation study. J. Clin. Med. **8**, 1865 (2019). https://doi.org/10.3390/jcm8111865
6. Hani, A.F.M., Kumar, D., Malik, A.S., Ahmad, R.M.K.R., Razak, R., Kiflie, A.: Non-invasive and in vivo assessment of osteoarthritic articular cartilage: a review on MRI investigations. Rheumatol. Int. **35**, 1–16 (2015). https://doi.org/10.1007/s00296-014-3052-9
7. Warnecke, D., et al.: Degeneration alters the biomechanical properties and structural composition of lateral human menisci. Osteoarthritis Cartilage **28**, 1482–1491 (2020). https://doi.org/10.1016/j.joca.2020.07.004
8. Desrochers, J., Amrein, M.W., Matyas, J.R.: Viscoelasticity of the articular cartilage surface in early osteoarthritis. Osteoarthritis Cartilage **20**, 413–421 (2012). https://doi.org/10.1016/j.joca.2012.01.011
9. Longo, U.G., Papalia, R., De Salvatore, S., Picozzi, R., Sarubbi, A., Denaro, V.: Induced models of osteoarthritis in animal models: a systematic review. Biology (Basel). **12**, 283 (2023). https://doi.org/10.3390/biology12020283
10. Moody, H.R., Brown, C.P., Bowden, J.C., Crawford, R.W., McElwain, D.L.S., Oloyede, A.O.: In vitro degradation of articular cartilage: Does trypsin treatment produce consistent results? J. Anat. **209**, 259–267 (2006). https://doi.org/10.1111/j.1469-7580.2006.00605.x

11. Roemer, F.W., Eckstein, F., Hayashi, D., Guermazi, A.: The role of imaging in osteoarthritis. Best Pract. Res. Clin. Rheumatol. **28**, 31–60 (2014). https://doi.org/10.1016/j.berh.2014.02.002
12. Joukar, A., et al.: Correlation analysis of cartilage wear with biochemical composition, viscoelastic properties and friction. J. Mech. Behav. Biomed. Mater. **142**, 105827 (2023). https://doi.org/10.1016/j.jmbbm.2023.105827
13. Hashim, N.H., Abd Latif, M.J., Jaafar, Y.L., Ramlan, R., Mahmud, J.: Computational and experimental study of articular cartilage thickness on biomechanical behavior. Int. J. Appl. Eng. Res. **12**, 5849–5856 (2017)
14. Fox, A.J.S., Bedi, A., Rodeo, S.A.: The basic science of articular cartilage: structure, composition, and function. Sports Health **1**, 461–468 (2009). https://doi.org/10.1177/1941738109350438
15. Wu, P.-J., et al.: Detection of proteoglycan loss from articular cartilage using Brillouin microscopy, with applications to osteoarthritis. Biomed. Opt. Express **10**, 2457–2466 (2019). https://doi.org/10.1364/boe.10.002457
16. Nia, H.T., et al.: High-bandwidth AFM-based rheology is a sensitive indicator of early cartilage aggrecan degradation relevant to mouse models of osteoarthritis. J. Biomech. **48**, 162–165 (2015). https://doi.org/10.1016/j.jbiomech.2014.11.012
17. Kabir, W., Di Bella, C., Choong, P.F.M., O'Connell, C.D.: Assessment of native human articular cartilage: a biomechanical protocol. Cartilage **13**, 427S-437S (2021). https://doi.org/10.1177/1947603520973240
18. Ibramsa, R.S., Latif, M.J.A., Zakaria, M.S., Harun, M.N., Mahmud, J.: Quantitative study of articular cartilage based on greyscale assessment using low-field MRI. Int. J. Adv. Sci. Eng. Inf. Technol. **12**, 231–236 (2022). https://doi.org/10.18517/ijaseit.12.1.13107
19. Temple, D.K., Cederlund, A.A., Lawless, B.M., Aspden, R.M., Espino, D.M.: Viscoelastic properties of human and bovine articular cartilage: a comparison of frequency-dependent trends. BMC Musculoskelet. Disord. **17**, 1–8 (2016). https://doi.org/10.1186/s12891-016-1279-1
20. Démarteau, O., Pillet, L., Inaebnit, A., Borens, O., Quinn, T.M.: Biomechanical characterization and in vitro mechanical injury of elderly human femoral head cartilage: comparison to adult bovine humeral head cartilage. Osteoarthritis Cartilage **14**, 589–596 (2006). https://doi.org/10.1016/j.joca.2005.12.011
21. Eschweiler, J., et al.: The biomechanics of cartilage-an overview. Life **11**, 302 (2021). https://doi.org/10.3390/life11040302

Comparative Preclinical Study of a Solid and a Porous Hip Stem for Canine Femora Using Finite Element Analysis

Soumyadeep Sarkar[1(✉)], Ram Niwas[2], and Souptick Chanda[1]

[1] Indian Institute of Technology, Guwahati, India
ssarkar@iitg.ac.in
[2] Lala Lajpat Rai University of Veterinary and Animal Sciences, Hisar, India

Abstract. The hip joint of heavier dogs gets flattened due to complications like hip dysplasia and osteoarthritis, leading to hip luxation. Total Hip Replacement (THR) is a common surgical treatment, in which artificial hip stems replace the natural hip joint. Although solid hip stems are majorly used, current research is shifting towards porous implants. The purpose of this study was to design a novel porous hip stem for canine femora and to run a comparative analysis with a solid hip stem. Triply periodic minimal surface (TPMS) gyroid structures were used to design the porous implant due to their light weight and ability to mimic the bone internal structure. In the finite element analysis (FEA), static compressive load was applied to the virtually implanted bone-stem models and the intact femur separately by mimicking the realistic loading conditions. Dynamic load was also applied to the hip stems based on gait cycle studies. Additionally, fatigue simulations were performed to assess the structural resilience of the implanted femur and stress shielding was evaluated. The porous implant showed lesser stress shielding as compared to the solid implant and was considered safe based on the simulation results. By assessing biomechanical parameters, the present study explored potential improvements in porous implants for canine femora.

Keywords: Canine hip joint · porous implant · stress shielding · triply periodic minimal surface · finite element analysis

1 Introduction

Heavy breeds of dogs, such as German Shepards and Golden Retrievers, suffer complications like hip dysplasia, capital physeal fractures, osteoarthritis, and hip luxation. In such conditions, the hip joint of these dogs, which is responsible for a smooth articulation between the acetabulum and the femoral head, suffers a major setback as one of these parts gets flattened due to the dog's overweight.

Total hip replacement (THR) is one of the available surgical options to remedy such complications, where the unstable hip joint is replaced by a new prosthesis. The THR typically consists of both the acetabular and the femoral components. The femoral stem is inserted into the femoral medullary canal to support the joint load and subsequently

C. T. Lim et al. (Eds.): ICBME 2024, IFMBE Proceedings 132, pp. 79–91, 2025.
https://doi.org/10.1007/978-3-032-04372-6_7

share it with the host bone. THRs are divided into two types: cemented and cementless. While the cemented stem achieves fixation via the cement mantle, the cementless femoral stems attain primary stability through the press fit mechanism and secondary stability biologically through osseointegration. Incorporating porous internal structure in such cementless femoral stems increases the potential for biologic fixation.

Stress shielding is known concern in THR where the metallic hip stem bears most of the joint load as opposed to the intact femur, which bears the entire load. This happens since the metallic implant possesses higher elastic modulus (100–200 GPa) compared to the bone (10–20 GPa). This imbalance in load sharing leads to a long-term decrease in bone density near the implanted hip stem, and subsequent loosening of the hip stem [1]. This phenomenon can be reduced significantly by utilizing porous scaffold structures, which can reduce the overall geometric stiffness of the hip stem. The scaffold structures, composed of repeating units of lattice unit cells, are used to design load-bearing implants [2, 3]. These structures mimic the bone microstructure and thus may allow the transport of nutrients and oxygen to the internal parts of the implant leading to osseointegration [4, 5].

Among the scaffold structures are the Triply Periodic Minimal Surface (TPMS) structures. TPMS unit cells are defined by sinusoidal functions [6]. The mechanical properties of TPMS structures can be tailored by adjusting their topology, which is further influenced by the unit cell type and the porosity. By modifying these properties, TPMS structures can be designed to match the mechanical characteristics of the respective bone they are replacing [7]. Porosity, which is determined by the ratio of material to void space, is controlled by varying the strut thickness in TPMS structures [8]. Due to their curved surfaces in all three directions, TPMS structures effectively eliminate localized stress concentrations [9]. They also exhibit a mean curvature value of zero, similar to human trabecular bone [10], which enhances their load-bearing capacity in implant applications [11]. Furthermore, a preclinical study by Wysocki et al. have demonstrated bone ingrowth within these structures, further supporting their suitability for implantation [12].

Prior to implementation of novel porous implants in surgery, computational studies are required to check the stability of the structure. Finite element analysis (FEA) is an acceptable tool to analyse the biomechanics of a bone-implant model by simulating realistic loading conditions in 3D models and checking their biomechanical stability. Thus, in this study, a novel porous hip stem was designed for canine femora and virtually implanted into a canine femur 3D model. The biomechanical characteristics of the porous hip stem were compared with a solid hip stem of similar dimensions using FEA.

2 Materials and Method

2.1 Design of the Solid and the Porous Hip Stems

A canine cadaveric femur was harvested from a male 2.5-year-old Golden Retriever followed by approval of the ethical committee. Segmentation of the Computed Tomography (CT) scan data of the canine femur was performed through medical image processing software (Mimics, Materialise, Belgium). A threshold was defined for the canine femur bones ranging from 700 to 2000 Hounsfield units (HU) [13]. Following segmentation, the femur 3D model was exported to Rhinoceros software (Rhinoceros®; McNeel, Seattle, USA) where the solid hip stem was designed using the femur as a reference. The porous hip stem was then designed using nTopology software (nTopology version 2.31), based on existing studies of scaffold structures (Fig. 1) [14, 15]. TPMS gyroid structure was used to design the porous section of the hip stem.

The TPMS gyroid scaffold structure was built using the following Eq. 1 [16]:

$$c = sin\frac{2\pi x}{L_x}.cos\frac{2\pi y}{L_y} + sin\frac{2\pi y}{L_y}.cos\frac{2\pi z}{L_z} + sin\frac{2\pi z}{L_z}.cos\frac{2\pi x}{L_x} \quad (1)$$

where x, y and z are number of cell repetitions in the x, y and z directions, L_x, L_y and L_z define the absolute size of the unit cells and c is the set level constant. A 50% porous TPMS Gyroid structure with a pore diameter of 600 μm was designed in this study, based on an existing study conducted by Barba et al. [17].

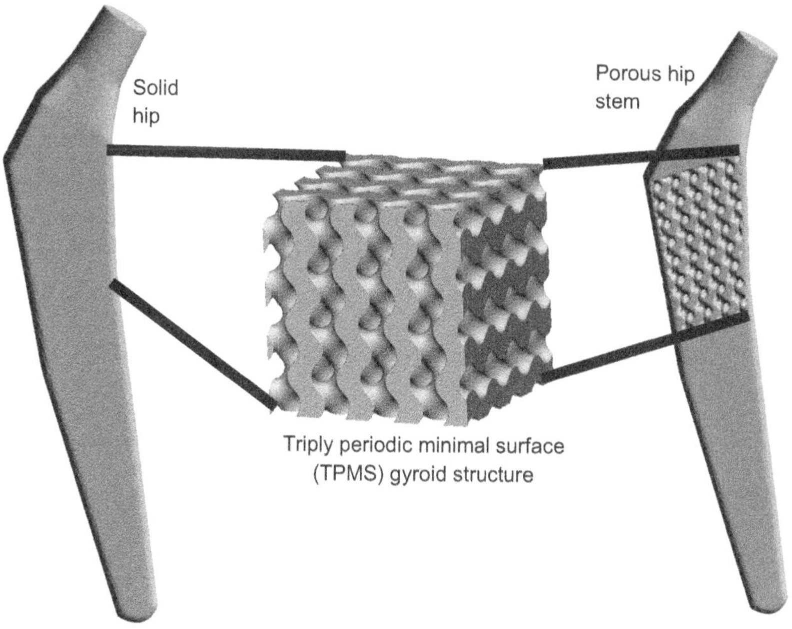

Fig. 1. Design of porous hip stem model by applying lattice structure to the solid hip stem model.

After designing, the hip stems were virtually implanted into the femur using Boolean subtraction operation and imported into Workbench (ANSYS, Inc., v.23.0) to perform FEA to assess the mechanical behavior of the bone-implant models (Fig. 2).

2.2 FEA – Mesh Convergence Study

Mesh convergence study was performed on the FE models prior to FEA (Table 1). Intact femur model, with realistic loading conditions of load at the femoral head and fixed

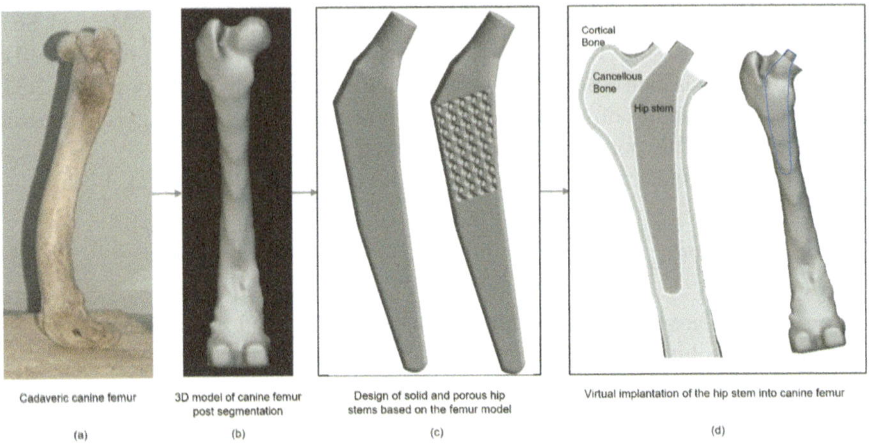

Fig. 2. Flow chart of the virtual implantation of hip stem into the canine femur: (a) a cadaveric canine femur was harvested, (b) which was segmented to obtain the 3D femur model, (c) the femur model was used as a reference to design the hip stem, and (d) the designed hip stem was then virtually implanted into the femur model.

support at the distal end of the femur was applied in the mesh convergence study, and the FE models were refined with mesh sizes ranging from 2 mm to 1 mm. Since the discrepancies were less than 5% for models 4–6, the model was considered to have converged at that point. Hence, model 6 (1 mm) was selected as the optimum mesh size for this FEA.

Table 1. Mesh convergence study details of the femur model

Model S. No.	Mesh element size (in mm)	Number of elements	Maximum Principal Stress (in MPa)	% Change in Maximum Stress
1	2	351657	98.17	-
2	1.8	366418	112	14.08
3	1.6	390771	79.62	28.91
4	1.4	431935	103.47	29.92
5	1.2	489838	100.45	2.9
6	1	594597	102	1.45

Table 2 summarizes the number of nodes and elements used in the analysis.

Titanium alloy Ti6Al4V being lightweight and biocompatible was considered as the material for the hip stem in this FEA. The FEA models were assumed to be linear, elastic and homogeneous. (Table 3) [18, 19]. The femur and the hip stem are assembled with a surface-to-surface contact with finite sliding, and a friction of coefficient of 0.4 is introduced to mimic the frictional contact in realistic conditions.

Table 2. FEA elements and nodes quantity

Geometry	Nodes	Elements
Cortical Bone (implanted with hip stem)	588541	336080
Cancellous Bone (implanted with hip stem)	498363	298967
Hip stem (solid)	138543	74700
Hip stem (porous)	378028	192878

Table 3. Mechanical material properties of the FE model components

Material	Young's Modulus (MPa)	Poisson's ratio	Yield Strength (MPa)
Titanium alloy	111200	0.33	845.7
Cortical Bone	16270	0.3	176
Cancellous Bone	660	0.3	80

2.3 FEA – Static Loading Analysis

For the first simulation, static compressive load was applied to the femoral head based on the maximal internal forces and moments acting in the femur during walking. The load was a factor of the bodyweight (BW) of the dog. In this FEA, the BW was assumed to be 300 N (Since the mass of the dog was 30 kgs). The contact force on the hip joint is 170% of the BW. Thus, the contact force at the hip corresponds to a load of 510 N to simulate peak joint reaction force. This load was applied on the femoral head of the intact bone and the head of the hip stem in the implanted bones respectively. The distal portion of the femur was given fixed support in all the models. (Fig. 3) For this FEA, the bones were abducted by 20° to mimic realistic loading conditions [20].

2.4 FEA – Dynamic Loading Analysis

Next, a dynamic compression simulation was conducted using Explicit Dynamics (ANSYS, Inc., v.23.0) on both the solid and the porous hip stems to check if the maximum stress observed in the implants under dynamic loading would be lower than the yield strength of the Ti6Al4V alloy used. To reduce the computational time of the simulation, only the hip stem models were chosen for this analysis, and the femur model was not included in the FEA. Following Sheikh et al. [21], the gait time was multiplied by a factor of 0.001, to reduce the CPU time required to execute the dynamic solution. The realistic dynamic load, calculated for the canine model used in this FEA [22, 23], was applied to the top of the respective hip stems for 1 s, while fixed support was provided to the distal portion of the hip stems (Fig. 4), following the ISO standards F2996 [24] and 7206-4 [25].

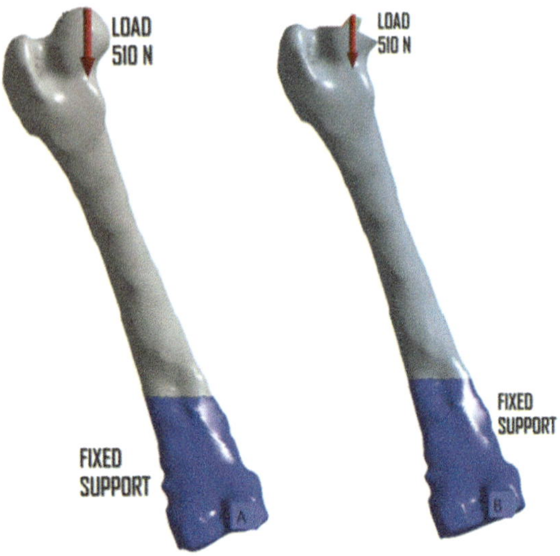

Fig. 3. Realistic loading and boundary conditions on (a) intact femur and (b) hip stem implanted femur.

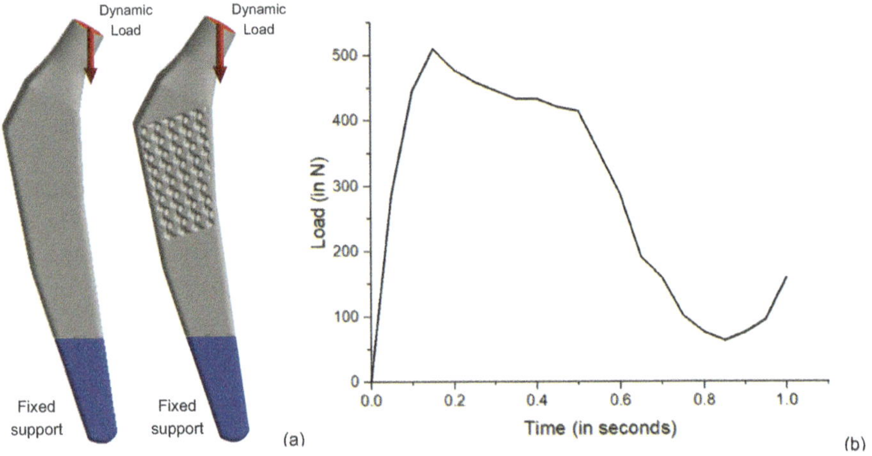

Fig. 4. (a) Loading and Boundary conditions of dynamic simulation using (b) time based realistic dynamic loading

2.5 FEA – Fatigue Analysis

A fatigue life analysis was conducted on both the hip stems. Cyclic load was applied to the head of the hip stems and fixed support was provided to the distal portion of the hip stems (Fig. 5). The maximum and minimum loads considered for the study were 510 N and 66 N, calculated for canine hip stems, from loads applied to human hip stems [26, 27]. Goodman's mean stress theory was used in this fatigue analysis [28]. The safety factor is calculated based on Eq. (1).

$$\frac{1}{SF} = \left(\frac{\sigma_a}{S_e}\right) + \left(\frac{\sigma_m}{S_u}\right) \tag{1}$$

where SF represents safety factor, σ_a and σ_m represent alternating stress and mean stress respectively, S_e and S_u represent endurance limit stress and ultimate strength respectively.

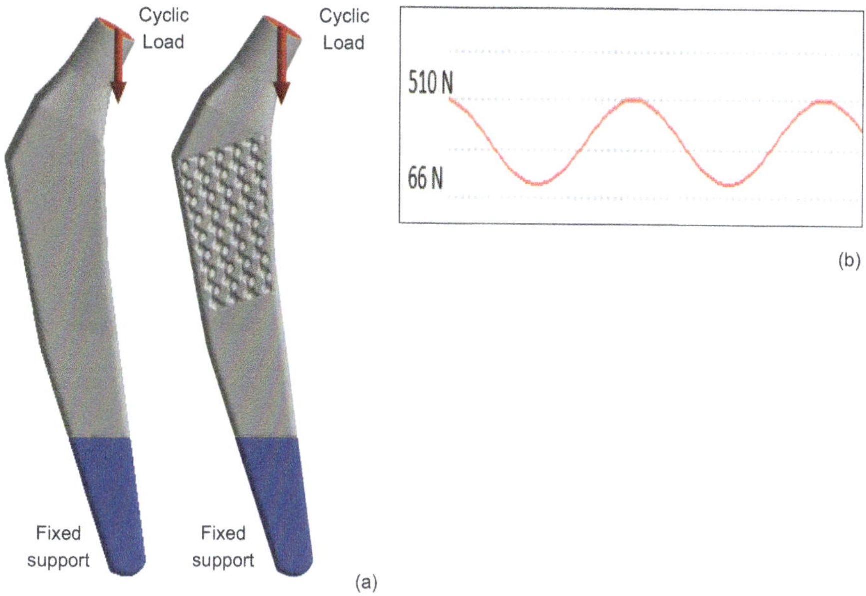

Fig. 5. (a) Loading and Boundary conditions of fatigue simulation using (b) cyclic loading

The damage or fatigue deterioration was calculated as the ratio between the reference number of cycles during the lifetime of a design and the predicted number of cycles to failure [29]. The designed hip stems were analyzed based on a predicted life cycle of 10 years where 1 year was equivalent to 1 million cycles [26]. 10 million cycles were considered as the life of the hip stem in this study.

3 Results and Discussion

In the static loading analysis conducted, the von Mises stress in the solid and porous hip stems and in the intact and hip stem implanted femurs were calculated and presented in Figs. 6, 7. From the stress contour of the hip stems (Fig. 6), it was observed that the stress was below 30 MPa throughout most of the solid hip stem. Stress of around 30–60 MPa was observed at the proximal end of the neck of the hip stem, directly below the line of point of load. The maximum observed stress was 300 MPa in the solid hip stem, which is far below the ultimate tensile strength of the Titanium alloy used. In the porous hip stem, in the scaffolded portion of the hip stem, the stress was around 60–120 MPa. Some porous portion of the hip stem showed high stress between 240 and 870 MPa, but the volume of such high stress zones was significantly low (0.3% volume of the total hip stem).

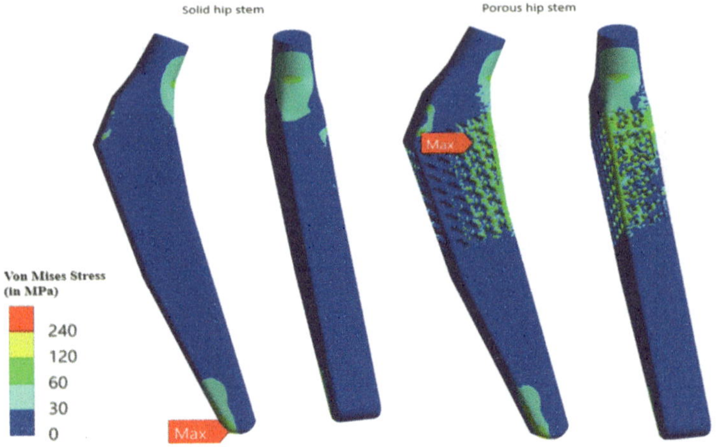

Fig. 6. Stress contour plot observed in the solid and porous hip stem

Next, the stress contour of the intact femur and the femur implanted with solid and porous hip stems respectively, have been compared (Fig. 7). The maximum yield strength of the cortical bone is 172.6 MPa and the stress transferred to the bone from the hip stems in this simulation were all well within the safety range of the stress that can be endured by the bone.

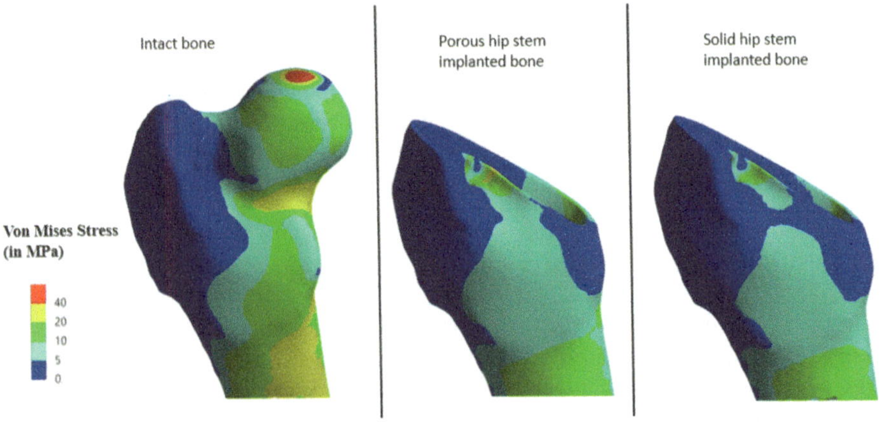

Fig. 7. Stress contour plot of proximal portion of the intact and implanted femur after static compression test

To calculate and compare the phenomenon of stress shielding, the average von Mises stress was calculated in the proximal portion of the canine femur for the intact bone and the bones implanted with hip stems.

Stress shielding is calculated from the following Eq. 2:

$$\text{Stress shielding (in \%)} = \left(\frac{\text{Difference in stress of intact bone and implanted bone}}{\text{Stress in intact bone}}\right) * 100 \tag{2}$$

Average stress observed in intact bone = 7 MPa.

Average stress observed in solid hip stem implanted bone = 4 MPa.

Average stress observed in porous hip stem implanted model = 4.9 MPa.

Stress shielding in solid hip stem = 42%

Stress shielding in porous hip stem = 30%

Thus, in the porous hip stem, stress shielding was significantly reduced compared to the solid hip stem.

The purpose of dynamic analysis was to check if the maximum stress observed in the solid and porous hip stem were within the yield strength of Titanium alloy Ti6Al4V used, i.e., 845.7 MPa. In the dynamic loading analysis conducted for both the hip stems, the maximum stress observed in one gait cycle was well under this value. The maximum stress observed in the solid hip stem was 392.5 MPa, while that observed in the porous hip stem was 521.2 MPa (Fig. 8).

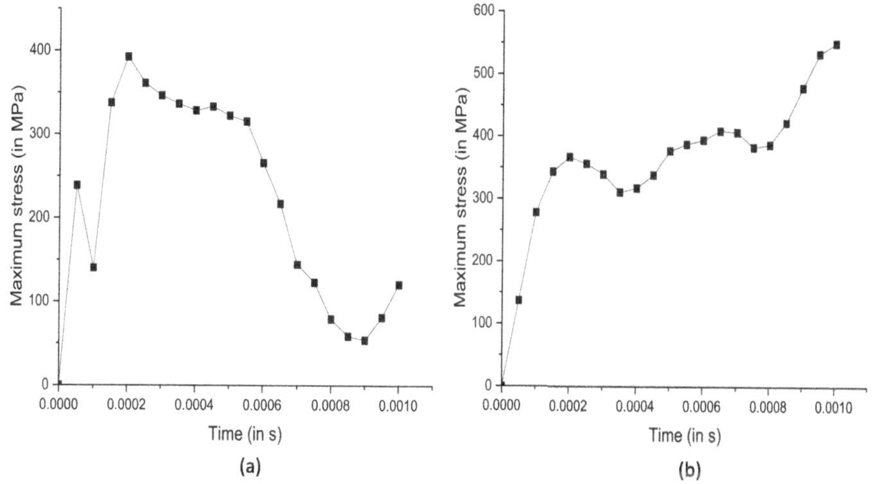

Fig. 8. Maximum stress observed in the (a) solid and (b) porous hip stems under the influence of dynamic loading

When the stress in a designed implant remains greatly inferior to the yield strength of the bulk material used, in this case, Ti-6Al-4V, the safety factor is higher than 1, rendering the model 'safe'. However, when the stress in a specific position becomes higher than the strength of the bulk Ti-6Al-4V, the safety factor ratio becomes lesser than 1, which means that in that specific area of the implant, the stress is higher than the strength that Ti-6Al-4V can bear and may ultimately fail. Damage or fatigue deterioration should ideally be between 0 and 1, as it is the ratio of the reference number of cycles during the lifetime of the hip implant (in this study, 10^7 cycles), and the predicted number of cycles to failure. s

In the fatigue analysis conducted in this study, both the solid and the porous hip stems are considered safe (Table 4), as per the considered life span of the hip stems i.e., 10 years.

Table 4. Fatigue results of the hip stem

Hip stems	Safety factor *(Safe range – 1 to 15)*	Damage *(Safe range – 0 to 1)*
Solid hip stem	14.63	0.33
Porous hip stem	12.62	0.17

In their study, Barba et al. discussed that the porosity levels of 50% and above, with pore sizes up to 600 μm in diameter, promote bone ingrowth, enhance osseointegration, and improve vascularization in additively manufactured structures [17]. Based on this, a 50% porous TPMS Gyroid structure with a pore diameter of 600 μm was selected in this study. Yanez et al. and Bobbert et al. reported that 50% porous TPMS Gyroid structures fabricated via SLM and EBM respectively, using Ti-6Al-4V, exhibited an elastic modulus of 5.2–5.6 GPa [30, 31]. Given that canine femoral trabecular bone has an elastic modulus ranging from 10 to 12 GPa [32], increasing the percentage of porosity more than 50% would have reduced the elastic modulus further, compromising the mechanical strength of the overall structure. Therefore, porous structures with a percentage of above 50% were not considered in this study, but will be explored in future research, similar to the study conducted by Bhardwaj et al. [33].

To ensure the clinical viability of the porous hip stem, follow-up studies are essential. These should include experimental testing on fabricated porous hip stems implanted in canine bones—either cadaveric or bone substitute models—followed by clinical validation in actual canine femora. A thorough investigation of osseointegration post-implantation is necessary to determine its potential application in veterinary medicine.

When transitioning from simulation to clinical use, several limitations must be considered. These include differences in the mechanical behavior of the porous hip stem between computational simulations and animal studies, potential failure of the porous section under extreme loading conditions unavailable in literature of veterinary hip stems, and the risk of implant subsidence due to inadequate primary stability and failed osseointegration.

In previous studies where the TPMS Gyroid structures were used in animal models, they exhibited a faster osteointegration compared to the conventional designs [34–36]. In rats, Van hede et al. demonstrated TPMS Gyroid scaffolds to have a better bone forming ability than an orthogonal lattice structure 3D-printed scaffold [34]. Mandibular bone defect was repaired using Gyroid structure in sheep in-vivo study [35]. In their study, Bouakaz et al. showed how the Gyroid structure was able to achieve ossification by mimicking a high percolation capacity similar to natural trabecular bones [36].

Thus, a successful animal study on canine femora would be a critical step toward confirming the sustainability of the designed porous implant for surgical application in veterinary medicine.

4 Conclusion

The study aimed to compare a novel canine porous hip stem with a solid hip stem using 3D FEA and analyze the stresses observed in the different hip stems and femur because of realistic loading conditions. Static and dynamic compressive loads were subjected at the femoral head of the intact and hip stem implanted canine femur 3D FE models. Additionally, a fatigue simulation was performed to check the stability of the hip stems in the long run. The results revealed that the porous hip stem reduced the stress shielding effects compared to the solid hip stem and was rendered safe to be used for at least 10 years. The porous hip stem was within the failure limits under both static and dynamic compressive loading. Although this study requires a clinical follow up study to assess the post-surgery bone osseointegration within the porous hip stem, it acts as a pilot study regarding the design and application of porous implants in animal models.

References

1. Savio, D., Bagno, A.: When the total hip replacement fails: a review on the stress-shielding effect. Processes **10**(3), 612 (2022)
2. Mullen, L., Stamp, R.C., Brooks, W.K., Jones, E., Sutcliffe, C.J.: Selective laser melting: a regular unit cell approach for the manufacture of porous, titanium, bone in-growth constructs, suitable for orthopedic applications. J Biomed Mater Res Part B Appl Biomater **89B**, 325–334 (2009)
3. Han, C., et al.: Effects of the unit cell topology on the compression properties of porous Co-Cr scaffolds fabricated via selective laser melting. Rapid Prototyp J **23**, 16–27 (2017)
4. Subasi, O., Karaismailoglu, B., Ashkani-Esfahani, S., Lazoglu, I.: Investigation of lattice infill parameters for additively manufactured bone fracture plates to reduce stress shielding. Comput. Biol. Med.. Biol. Med. **161**, 107062 (2023)
5. Al Zoubi, N.F., Tarlochan, F., Mehboob, H., Jarrar, F.: Design of titanium alloy femoral stem cellular structure for stress shielding and stem stability: computational analysis. Appl. Sci. **12**, 1548 (2022)
6. Karcher, H.: The triply periodic minimal surfaces of alan schoen and their constant mean curvature companions. Manuscripta Math. Math. **64**, 291–357 (1989)
7. Maconachie, T., et al.: SLM lattice structures: properties, performance, applications and challenges. Mater. Des. **183**, 108137 (2019)
8. Wall, A., Board, T.: The Compressive Behavior of Bone as a Two-phase Porous Structure, Classic Papers in Orthopaedics. Springer, pp. 457–460 (2014)
9. Hameed, P., Liu, C.-F., Ummethala, R., Singh, N., Huang, H.-H., Manivasagam, G., Prashanth, K.G.: Biomorphic porous Ti6Al4V Gyroid scaffolds for bone implant applications fabricated by selective laser melting. Progress Addit. Manuf. **6**(3), 455–469 (2021)

10. Bobbert, F.S.L., Zadpoor, A.A.: Effects of bone substitute architecture and surface properties on cell response, angiogenesis, and structure of new bone. J Mater Chem B **5**, 6175–6192 (2017)
11. Zadpoor, A.A.: Bone tissue regeneration: The role of scaffold geometry. Biomater. Sci. **3**, 231–245 (2015)
12. Wysocki, B., et al.: Post processing and biological evaluation of the titanium scaffolds for bone tissue engineering. Materials **9**(3), 197 (2016)
13. Saunders, W.B., et al.: Angle-stable interlocking nailing in a canine critical-sized femoral defect model for bone regeneration studies: In pursuit of the principle of the 3R's. Front. Bioeng. Biotechnol. **10**, 921486 (2022)
14. Salaha, Z.F.M., Ammarullah, M.I., Abdullah, N.N.A.A., Aziz, A.U.A., Gan, H.-S., Abdullah, A.H., Abdul Kadir, M.R., Ramlee, M.H.: Biomechanical effects of the porous structure of Gyroid and Voronoi Hip implants: a finite element analysis using an experimentally validated model. Materials **16**, 3298 (2023)
15. Blanquer, S.B.G., Werner, M., Hannula, M., Sharifi, S., Lajoinie, G.P.R., Eglin, D., Hyttinen, J., Poot, A.A., Grijpma, D.W.: Surface curvature in triply-periodic minimal surface architectures as a distinct design parameter in preparing advanced tissue engineering scaffolds. Biofabrication **9**(2), 025001 (2017)
16. Wallat, L., Altschuh, P., Reder, M., Nestler, B., Poehler, F.: Computational design and characterisation of gyroid structures with different gradient functions for porosity adjustment. Materials **15**, 3730 (2022)
17. Barba, D., Alabort, E., Reed, R.C.: Synthetic bone: design by additive manufacturing. Acta Biomater. Biomater. **97**, 637–656 (2019)
18. Vahey, J.W., Lewis, J.L., Vanderby, R.: Elastic moduli, yield stress, and ultimate stress of cancellous bone in the canine proximal femur. J. Biomech.Biomech. **20**(1), 29–33 (1987)
19. Autefage, A., Palierne, S., Charron, C., Swider, P.: Effective mechanical properties of diaphyseal cortical bone in the canine femur. Vet. J. **194**(2), 202–209 (2012)
20. Heo, S., Lee, H., Roh, Y., Jeong, J.: Biomechanical comparison between inverted triangle and vertical configurations of three Kirschner Wires for Femoral neck fracture fixation in dogs: a cadaveric study. Vet. Sci. **10**(4), 285 (2023)
21. El'Sheikh, H.F., MacDonald, B.J., Hashmi, M.S.: Finite element simulation of the hip joint during stumbling: a comparison between static and dynamic loading. J. Mater. Process. Technol. **143**, 249–255 (2003)
22. Kuo, T.Y., Skedros, J.G., Bloebaum, R.D.: Comparison of human, primate, and canine femora: Implications for biomaterials testing in total hip replacement. J. Biomed. Mater. Res. **40**, 475–489 (1998)
23. Bergmann, G., et al.: Realistic loads for testing hip implants. Biomed. Mater. Eng. **20**(2), 65–75 (2010)
24. ASTM. F 2996-20 Standard Practice for Finite Element Analysis (FEA) of Non-Modular Metallic Orthopaedic Hip Femoral Stems. ASTM International, West Conshohocken, PA, 2020:1–11. https://doi.org/10.1520/F2996-13.2 last Accessed 15 Oct 2024
25. ISO 7206-4:2010 - Implants for surgery—Partial and total hip joint prostheses—Part 4: Determination of endurance properties and performance of stemmed femoral components. https://www.iso.org/standard/42769.html last Accessed 15 Oct 2024
26. Senalp, A., Kayabasi, O., Kurtaran, H.: Static, dynamic and fatigue behavior of newly designed stem shapes for hip prosthesis using finite element analysis. Mater. Des. **28**(5), 1577–1583 (2007)
27. Delikanli, Y.: Design, manufacture, and fatigue analysis of lightweight hip implants. J. Appl. Biomater. Funct. Mater. 17(2), (2019)

28. Aghili, S., Hassani, K., Nikkhoo, M.: A finite element study of fatigue load effects on total hip joint prosthesis. Comput. Methods Biomech. Biomed. Engin.. Methods Biomech. Biomed. Engin. **24**, 1545–1551 (2021)
29. Han, C., Qu, X., Ma, Y., Shi, D.: Experimental and numerical study of fatigue damage assessment under combined high and low cycle loading. Shock and Vibration 1–12 (2018)
30. Yanez, A.: Gyroid porous titanium structures: a versatile solution to be used as scaffolds in bone defect reconstruction. Mater. Des. **140**, 21–29 (2018)
31. Bobbert, F.: Additively manufactured metallic porous biomaterials based on minimal surfaces: a unique combination of topological, mechanical, and mass transport properties. Acta Biomater. Biomater. **53**, 572–584 (2017)
32. Pressel, T., Bouguecha, A., Vogt, U.: Mechanical properties of femoral trabecular bone in dogs. BioMed Eng **4**, 17 (2005)
33. Bhardwaj, J.S., Chanda, S.: Structural and hemodynamic analysis of Weaire-Phelan scaffolds made of Ti-alloy as bone replacement component: a preclinical investigation. PLoS ONE **19**(12), e0312880 (2024)
34. Van hede, D., Liang, B., Anania, S., Barzegari, M., Verlée, B., Nolens, G., Pirson, J., Geris, L., Lambert, F.: 3D-printed synthetic hydroxyapatite scaffold with in silico optimized macrostructure enhances bone formation in vivo. Adv. Funct. Mater. **32**(6) (2022)
35. Paré, A., et al.: Tailored three-dimensionally printed triply periodic calcium phosphate implants: a preclinical study for craniofacial bone repair. ACS Bio-mater. Sci. Eng. **6**(1), 553–563 (2020)
36. Bouakaz, I., Drouet, C., Grossin, D., Cobraiville, E., Nolens, G.: Hydroxyapatite 3D-printed scaffolds with Gyroid-Triply periodic minimal surface (TPMS) porous structure: fabrication and an in vivo pilot study in sheep. Acta Biomater. Biomater. **170**, 580–595 (2023)

Design and Development of a Highly Sensitive Aptamer Based Electrochemical Sensor Using Reduced Graphene Oxide–Gold Nanocomposite for Detection of Enrofloxacin

Deepak Saravanan and Monisha Mohan(✉)

School of Interdisciplinary Design and Innovation, Indian Institute of Information Technology, Design and Manufacturing, Kancheepuram, Chennai, Tamil Nadu 600127, India
monisha@iiitdm.ac.in

Abstract. Enrofloxacin, a fluoroquinolone widely used in animal husbandry, poses significant risks to human health and the environment due to residue accumulation and antibiotic resistance. Aptamer-based electrochemical sensors provide the real-time measurement of the biomolecules from the samples, making them a powerful technique in clinical and research applications in recent years. This study presents the development of a novel, label-free electrochemical aptasensor for the sensitive and selective detection of enrofloxacin. The sensor utilizes a reduced graphene oxide–gold nanoparticle (rGO–AuNP) composite, which provides a high surface area, enhanced electron transfer, and excellent conductivity for aptamer immobilization. The rGO-AuNP nanocomposite was characterized using UV-Vis spectroscopy, scanning electron microscopy, X-ray diffraction, and Fourier transform infrared spectroscopy, while its electrochemical properties were evaluated using cyclic voltammetry and differential pulse voltammetry. The recovery rate of enrofloxacin residues in milk samples was 92.11–115.51%. Demonstrating its potential applications in food safety control. The aptasensor exhibited high sensitivity, selectivity, and repeatability for enrofloxacin detection, with minimal interference from other antibiotics. Its performance highlights its potential as a cost-effective and efficient tool for food safety applications.

Keywords: Enrofloxacin · reduced Graphene Oxide and Gold nanocomposite · Electrochemical analysis · Modified electrode · Peak current

1 Introduction

In recent years, antibiotics have played a vital role in the treatment of various ailments across the human as well as veterinary medicine [1]. Among the antibiotics, Fluoroquinolone, which has a wide range of antibacterial activity for both the human and animals, treats the infection in the respiratory, gastrointestinal, and urinary tract [2, 3]. Fluoroquinolone antibiotics are generally supplemented in the water bodies to eradicate the propagation of microbes and also prevent the aquatic products from rotting [4, 5]. Enrofloxacin is an antibiotic of the fluoroquinolone family used as the medication for

C. T. Lim et al. (Eds.): ICBME 2024, IFMBE Proceedings 132, pp. 92–105, 2025.
https://doi.org/10.1007/978-3-032-04372-6_8

the animals to cure infection or to enhance the growth and productivity [6]. However, enhanced usage of enrofloxacin resulted in antibiotic resistance to various microorganisms [7]. The enhanced level of enrofloxacin in the aquatic foods poses high risk to human health [8]. Therefore, many countries across the globe have set the maximum residue limits (MRLs) for the antibiotics in food derived from animals. Currently, the MRLs of enrofloxacin ranges from 100 μg/kg to 300 μg/kg depending on the nature of the animal source and their tissue type [9]. Early times, the enrofloxacin was detected using various analytical methods such as liquid chromatography combined with mass spectrometry (LC-MS), high-performance liquid chromatography (HPLC) [10], HPLC coupled with ultraviolet detect (HPLC-UV) [11] and fluorescence detection (HPLC-FLD) [12], enzyme-linked immunosorbent assay (ELISA) [13], microbial methods [14], immunochromatographic test strips [14, 15], and the molecular imprinting techniques [16]. These approaches have proven to have high sensitivity and accuracy, meanwhile they have disadvantages in terms of time consumption, cost, complex operating protocols, large workstation, and well trained operators which affects the on-site detection [16, 17].

Gold nanoparticles (AuNPs) are biologically friendly, have good conductivity, and can improve the reactivity and durability of graphene-modified electrodes due to their unique surface, chemically inactivity, high density of electrons, and have significant optical absorption. Gold nanoparticles are currently being employed in genomics [18], clinical chemistry [19], drug development [20], biosensors [21], immunoassay [22], diagnostics and microorganism control [23], bio imaging of cancer cells [24], and energy storage [25]. The combination of AuNPs with rGO has resulted in several opportunities and applications due to the component elements' synergistic effects, potentially expanding graphene's application range. An atomically thin sheet of carbon is called reduced graphene oxide (rGO) [26] which is a truncated form of graphene oxide is widely used as a support material in electrochemical applications due to its strong electro catalytic activities, large surface area-to-volume ratio, and intriguing electrical transport properties [27]. Because of this, rGO was incorporated with metal nanoparticles that have been used in concert to detect environmental pollutants in electrochemical sensors [28–30].

Aptamers are artificial single-stranded DNA or RNA oligonucleotides (10–100 bases) chosen using exponential enrichment (SELEX) [31]. Aptamers effectively bind to targets by folding into precise spatial conformations. Aptamers are widely employed in biological analysis [32], pharmaceutical [33], and food safety analysis [34] due to their low cost, high affinity and specificity, ease of modification, and long-term storage. Suozhu Wu et al. proposed a nanocomposite based electrochemical sensor based on reduced graphene oxide coupled with carboxylated multi-walled carbon nanotubes for the enrofloxacin detection [35]. Amed Gallegos-Tabanico et al. developed an electrochemical sensor based on Molecular Imprints Polymer (MIP) coated on the modified screen printed gold-electrodes for the detection of enrofloxacin [36]. Zijiu Niu et al. proposed a aptamer based biosensor for enrofloxacin detection from a liver of sheep and pork by a bimetallic carbide of cobalt and nickel nanostructures [37]. Naveen Karuppusamy et al. developed an electrochemical based sensor by incorporation of reduced graphene oxide with tungsten metal for the enrofloxacin detection [38]. As a result of previous studies, there is an urgent need to create a highly sensitive, fast, convenient, and dependable approach for detecting enrofloxacin metabolites. In this paper, a rGO-AuNPs composite-based aptamer biosensor was produced, characterized, and employed for the first time to detect enrofloxacin. The proposed electrode appears promising since it enables the rapid, simple, low-cost, and sensitive detection of enrofloxacin in aquatic foods.

2 Materials and Methods

2.1 Reagents

Reduced Graphene oxide, Gold(III) chloride tetrahydrate, TCEP (Tris(2-carboxyethyl) phosphine hydrochloride), Trisodium citrate, Sodium borate, Ethanol, and Distilled water. The aptamer with the following sequence: 5‘-CCCATCAGGGGGCTAGGCT AACACGGTTCGGCTCTCTGAGCCCGGGTTATTTCAGGGGGA-3‘ was obtained from sigma [39]. Enrofloxacin was purchased from Sigma Company (Sisco Research Laboratories PVT LTD).

2.2 Instruments

Fourier transform infrared (FTIR) spectra were recorded using SHIMADZU, IRTRACER 100 in the range of 400–4000 cm − 1. rGO and Au/rGO NCs were characterized by X-Ray Diffraction technique (XRD) (9 kW) - RIGAKU, JAPAN. The morphologies of the samples were characterized using Field emission scanning electron microscopy (FESEM) - CARL ZEISS, GERMANY. Electrochemical analysis was performed at room temperature using an electrochemical analyzer Squidstat solo (Potentiostat).

2.3 Preparation rGO-Au Nanocomposite

The synthesis of rGO/Au nanocomposite was done by *in situ* production of Gold nanoparticles coating onto the surface of Graphene (Fig. 1). The stock solution of gold chloride tetrahydrate was prepared by adding 1 ml of Molecular grade water to 1 g of $AuCl_3H_8O_4$ and stored at 4 °C in a dark environment. 20 mg of reduced graphene oxide was added to the $AuCl_3H_8O_4$ suspension, and made up of volume to 2 mL by molecular grade water and subjected to sonication for 30 min at room temperature. After sonication, 100 mg of trisodium citrate and 200 mg of NaBH4 was dissolved in 8 ml of distilled water and added to the sonicated mixture. The resulting mixture was allowed to stir for 6 h at 70 °C at 600 rpm. The nanocomposite was washed thrice with ethanol and the precipitated slurry was dried in a hot air oven at 50 °C overnight.

2.4 Characterization of rGO-Au Nanocomposite

The characterization of the prepared nanocomposite was carried out by different techniques. Scanning electron microscopy (SEM) (Carl Zeiss NTS GMBH, Germany, SUPRA 55) was used to study the size and morphology of rGO-AuNP. The hydrodynamic diameter of the particle was assessed using Zeta Nano ZS90. The crystalline structure of the rGO-Au NPs was studied by X-ray Diffraction (XRD) (PANalytical, Netherlands) which was supplied with the Cu-Kα Monochromatic source and Ni filter. Also, Fourier Transform Infrared Radiation (FTIR) spectrum (SHIMADZU, IRTRACER 100) was used to investigate the functional group of rGO-AuNP. The Raman spectra was investigated using Wasatch Photonics, USA with excitation of 632 nm.

2.5 Pretreatment of Glassy Carbon Electrode

The pretreatment of GCE was done by Cyclic voltammetry. Electrochemical analysis was performed with potentiostat device using three electrode system comprising of a Working electrode namely; 3 mm diameter glassy carbon electrode, a reference electrode (Ag/AgCl), and a counter electrode (platinum) At first, the GCE was prepared by successive polishing with the following alumina powder which has the size of 1.0, 0.3, and 0.05μm on a micro cloth. The polished electrode was subjected to ultra-sonication and washed with distilled water to remove the unbounded particles. The unfunctionalized electrode was activated by cyclic voltammetry analysis at scan rate of 10 mv/s in the range of − 1V to 1V for up to 10 cycles in 1x PBS buffer containing 0.1M H_2O_2. After each electrochemical analysis, the electrode was cleaned and activated for further studies.

2.6 Fabrication of the Sensor

For the fabrication of the sensor, 10 μL of rGO-Au nanoparticle suspension was drop casted on the bare electrode and allowed it to dry at room temperature for 2 h. To functionalize the electrode with the aptamer, 200 μL of TCEP (1 mmol) was mixed with 20 μL of the aptamer (2 μM) and incubated at 37 °C for 1 h. After incubation, 10 μL of this aptamer mixture was drop casted onto the rGO-Au coated electrode, which was then incubated overnight at 4 °C, completing the sensor fabrication process.

2.7 Electrochemical Measurement

All the electrochemical experiment was conducted using a Squidstat Solo Potentiostat workstation, employing a three-electrode setup. The electrolyte for these experiments consisted of a 20 mM solution of $\left[Fe(CN)_6\right]^{4-}/^{3-}$ in 0.1 M KCl, providing a stable environment for redox reactions. The modified glassy carbon electrode, aptamer-coated rGO-Au composite, was optimized for the effect of pH, aptamer concentration, and incubation time to ensure the accurate detection of enrofloxacin. To test the specificity of the sensor, cyclic voltammetry (CV) and the Differential pulse voltammetry (DPV) analysis was performed with different antibiotic such as Doxycycline hydrate (DHX), and Tetracycline hydrate (TC).

3 Results and Discussion

3.1 Characterization of rGO-Au Nanocomposite

X-ray diffraction of rGO-Au nanocomposite (Fig. 2) shows that diffraction plane peaks of Au (111), Au (200), Au (220), Au (311), and Au (222), and corresponding to 2θ angle are 38.00°, 44.48°, 64.4°, 77.6°, and 81.5°, respectively, showing face-centered cubic (FCC) crystal structure. A significant (111) diffraction peak suggests a prominent growth of network structure along (111) planes compared to (200). Additionally, the other prominent diffraction peak (220) points to the anisotropic (network) nature of the nanoparticles. This is consistent with diffraction peak position of JCPDS database (Au, JCPDS file: 04-0784). It confirms that AuNPs are adsorbed onto the surface of the graphene layer (Fig. 1A). The surface functional groups of graphene gold nanocomposite were investigated by ATR-FTIR spectroscopy (Fig. 3). The reduction of GO was depicted base on the weak peaks of –OH, C=O and O–H Bond vibrations situated at 3684, 1571, And 1217 cm^{-1} respectively. In the case of the functionalized and reduced graphene oxide, the intensity of these peaks decreased dramatically or even disappeared. It also observed the additional peaks at 2160.27, 2027.87, and 1973.17 cm^{-1} (Fig. 1B).

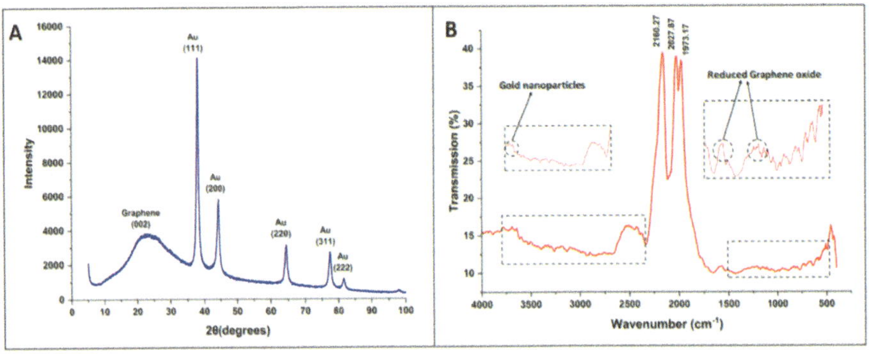

Fig. 1. Characterization of rGO-Au nanocomposite. (A) XRD, and (B) FTIR.

The morphology of AuNP/rGO samples was characterized by FE-SEM technique. As it can be easily seen, the structure strongly depends on the drying method used. After simultaneous reduction of GO in the presence of HAuCl4, the surface of rGO nanosheets was decorated with AuNPs (Fig. 2A and B). Furthermore, elemental analysis further confirmed the presence of gold on the reduced graphene oxide surface (Fig. 2C). Raman spectroscopy was employed to characterize the structural properties of the rGO-AuNP nanocomposite, revealing significant insights into its composition and quality. The analysis displayed distinct peaks corresponding to the D band at approximately 1312.36 cm $^{-1}$ and the G band at around 1577 cm^{-1} (Fig. 2D). The D band corresponds to the breathing mode in sp^2-hybridized carbon rings, with increased intensity indicating a greater presence of sp^2 domains and the presence of double bonds (C=C). The G band, associated with the E2g phonon mode of sp^2 carbon atoms, reflects first-order phonon dispersion. The ratio of the D to G band intensities (ID/IG = 1.309) and the absence of the 2D band support the formation of reduced graphene oxide (rGO). Dynamic Light Scattering (DLS) was performed at a scattering angle of $\theta = 90°$ to characterize the aqueous dispersions of the rGO-AuNP. In DLS, the conventional spherical particle models were employed. rGO-AuNP had a hydrodynamic diameter of 668 ± 163.8 nm (Fig. 2E). The model-derived diameters are not the actual sizes of the majority of graphene-based materials because they are not spherical particles. The rGO-Au NPs' zeta potential was – 21.4 mV, indicating the composite's strong stability (Fig. 2F).

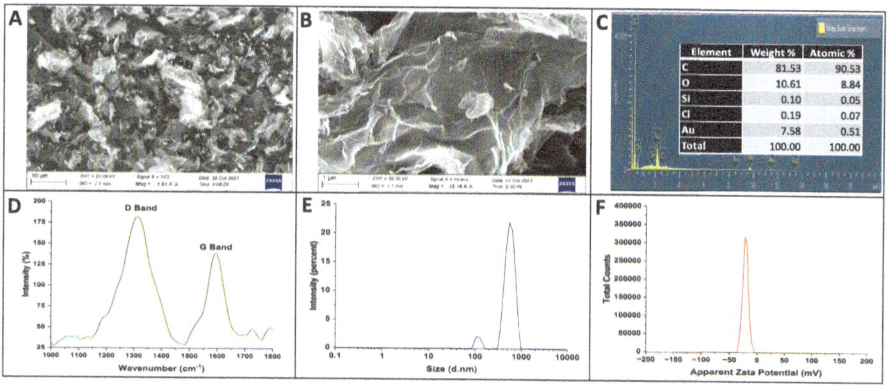

Fig. 2. Characterization of rGO-Au nanocomposite. SEM images showing Au nanoparticles uniformly distributed across the rGO sheets (A and B). EDS spectrum confirms the elemental composition (C). Raman spectroscopy analysis displaying an ID/IG ratio of 1.309, confirming successful reduction of GO (D). DLS measurements illustrate the hydrodynamic diameter of the rGO-Au NPs (E). Zeta potential analysis shows a value of – 21.4 mV, indicating good colloidal stability of the composite (F).

3.2 Electrochemical Performance of the Electrode

To investigate the electrochemical sensitivity of rGO-Au/apt modified GCE for detecting ENF, CV measurement was performed on bare GCE, GCE/rGO-Au, GCE/rGO-Au/apt and GCE/rGO-Au/apt/ENF (Fig. 3A). Upon addition of ENF to rGO-Au/apt, the peak current decreased to 72.5 μA from 80 μA, suggesting an increase in resistance due to ENF binding with the aptamer. Additionally, CV measurements across varying scan rates showed that the anodic and cathodic peak currents increased proportionally with scan rate, confirming a linear relationship between current and scan rate for the modified electrode (Fig. 3B and C). This linearity in current (I) with scan rate (X) for the modified electrode can be expressed as $I = 0.8206^{*} X + 88.85$.

3.3 Optimization of Working Condition

3.3.1 Effect of Aptamer Concentration

Optimizing aptamer concentration on the modified electrode is essential for detecting ENF effectively. Aptamer concentrations from 0.05 to 3 μM were tested to determine the optimal binding level (Fig. 4). ENF detection was carried out at a fixed concentration of 1μM. Results indicated that peak current increased progressively with aptamer concentration, reaching a maximum at 0.25 μM of aptamer, as observed by DPV. Thus, 0.25 μM was identified as the optimal aptamer concentration for aptasensor fabrication, producing a peak current of 23.20 MA.

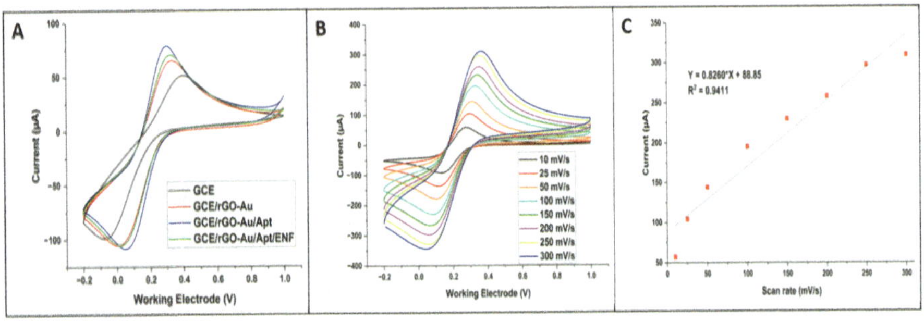

Fig. 3. Cyclic voltammogram of bare GCE and GCE/rGO-Au nanocomposite in 0.1 M KCl containing 20 mM $\left[Fe(CN)_6\right]^{4-}/^{3-}$ (A). CV of the GCE/rGO-Au sensor at various scan rates in the same solution (B). Linear plot of peak current versus scan rate, demonstrating a surface-controlled electrochemical process (C).

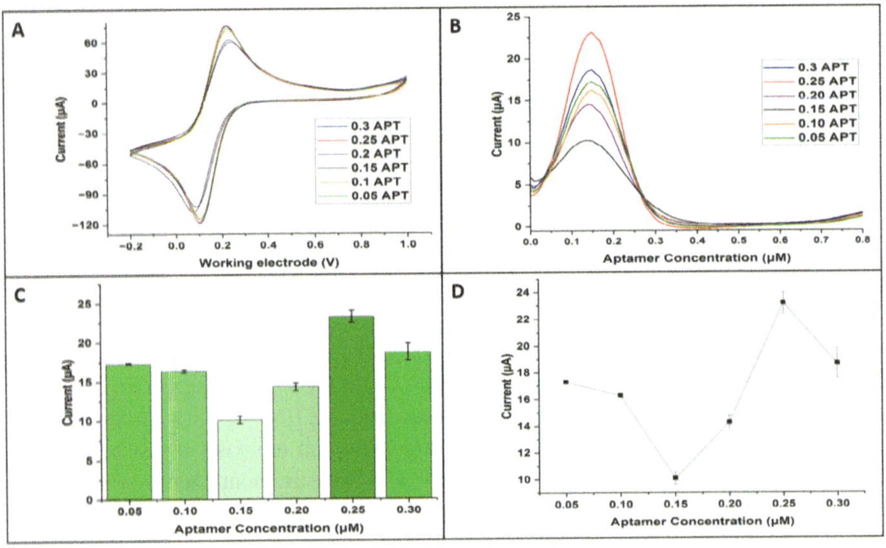

Fig. 4. Optimization of aptamer concentration. CV for varying aptamer concentrations (0.05–3 μM) in 0.1 M KCl containing 20 mM $K_4\left[Fe(CN)_6\right]^{4-}/^{3-}$ (A). DPV results highlight the peak currents at different aptamer concentrations (B). Bar graph and line graph illustrating the peak current response across the tested aptamer concentrations (C & D).

3.3.2 Effect of pH

The pH of the electrolyte is a crucial parameter in sensing ENF, as it influences electron transfer during electrochemical analysis. The effects of varying pH levels (3 to 9) of a 0.1 M KCl solution containing 20 mM $\left[Fe(CN)_6\right]^{4-}/^{3-}$ (potassium ferricyanide) were investigated (Fig. 5). DPV results showed that the peak current decreased with increasing pH, with the maximum response observed at pH 3. The downward trend in peak current with increasing pH indicates that a basic environment significantly affects the performance of the aptasensor.

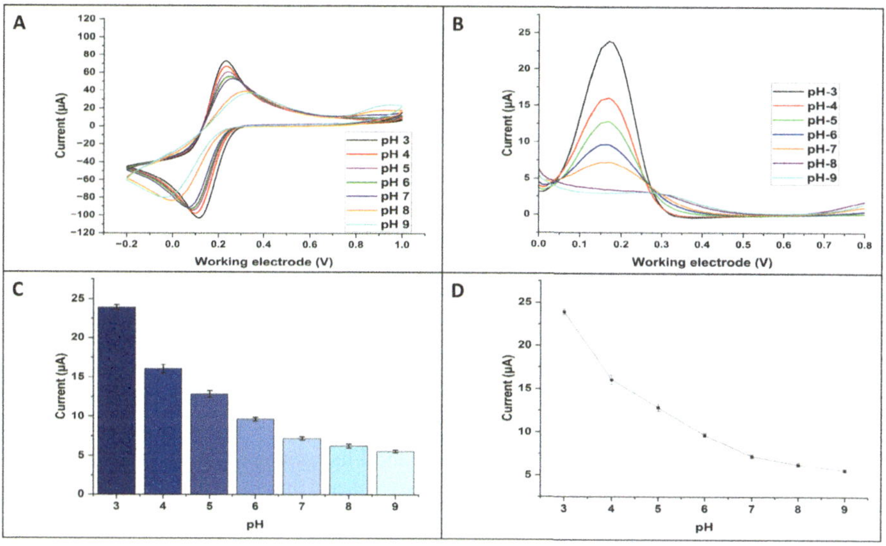

Fig. 5. Optimization of pH for ENF detection. DPV results for various pH levels (3–9) in 0.1 M KCl containing 20 mM $K_4[Fe(CN)_6]$ (A). CV illustrating the electrochemical response at different pH values (B). Bar graph and Line plot depicting the relationship between pH and peak current (C & D).

3.3.3 Effect of Incubation Time

To determine the optimal incubation time for ENF sensing, a modified electrode was immersed in a potassium ferricyanide electrolyte solution with a 1 μM ENF concentration. Using DPV to monitor sensor response, the maximum current was observed at 0 min, followed by a gradual decrease, stabilizing at approximately 20 min (Fig. 6). This stabilization suggests that equilibrium is reached between enrofloxacin and its aptamers. Therefore, an incubation time of 20 min is identified as optimal for ENF detection.

3.3.4 Effect of Enrofloxacin Concentration

To elucidate the analytical performance of the designed aptasensor, experiments were carried out by adding different concentrations of enrofloxacin to the aptasensor. The different concentrations of enrofloxacin (0.05 to 1.5 μM) in the 0.1M KCl containing 20 mM $[Fe(CN)_6]^{4-}/^{3-}$ (potassium ferricyanide) electrolyte were investigated. Under the optimal experimental conditions, the CV and DPV peak current decreased significantly with the increasing concentration of the target ENF (Fig. 7). This reduction in the current upon addition of ENF confirms a strong interaction between the aptasensor and the target. The LOD of GCE/rGO-Au/apt sensor detecting Enrofloxacin is 0.2613 μM with range of detection of 0.05 to 1 μM, as shown in Table 1.

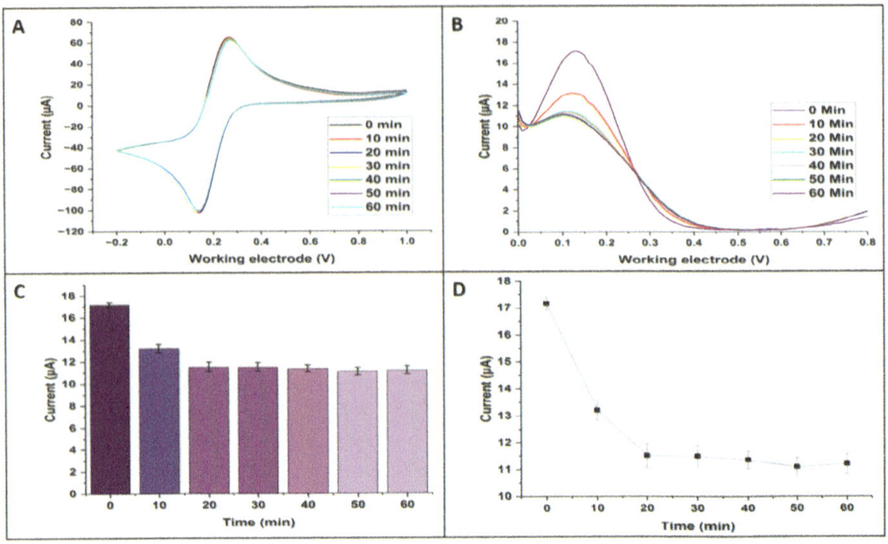

Fig. 6. Optimization of Sensing Period. Electrochemical measurements were conducted to optimize incubation time using CV (A) and DPV (B). Bar graph and line graph displaying peak current response trends across incubation times (C&D).

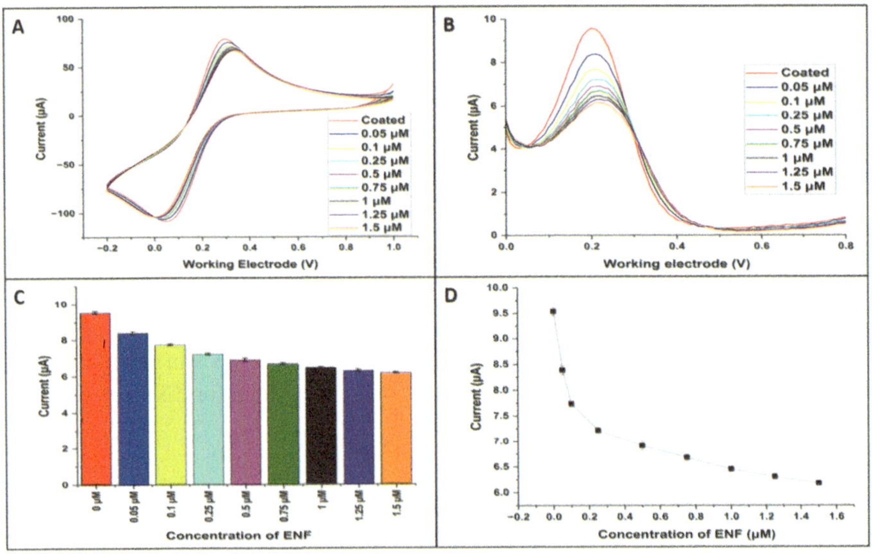

Fig. 7. Electrochemical measurements evaluating the effect of varying ENF concentration. CV (A) and DPV (B) responses illustrating peak current variations at different ENF concentrations. Bar graph and line plot showing current response trends across varying ENF concentrations (C and D).

Table 1. Comparison of antibiotic detection by electrochemical analysis

Antibiotic	Sensing platform	Detection Limit (μM)	Range of detection (μM)	Ref.
Tetracycline	PtNPs/C/GCE	4.28	9.99 to 44.01	[40]
Sulfamethoxazole	TYR-AuNPs/SPCE	22.6	20 to 200	[41]
Tetracycline	AuNPs@MoS_2 /Ch	0.41	1 to 1000	[42]
Enrofloxacin	MIP	20	10^4 to 10^5	[36]
Ciprofloxacin	Au/C3N4/GN/GCE	0.42	0.6 to 120	[43]
Chloramphenicol	Au/N-G	0.59	2 to 80	[44]
Levofloxacin	MIP/G-AuNPs	0.53	1 to 100	[45]
Chlortetracycline	CS@MnO2/GCE	0.26	0.5 to 300	[46]
Enrofloxacin	rGO-Au/apt	0.26	0.05 to 1	Present work

3.4 Specificity Analysis for the Detection of Enrofloxacin

To assess the selectivity of the aptasensor, its response to various antibiotics was examined using DPV. The specificity of enrofloxacin (ENF) towards the modified electrode was analyzed with Ciprofloxacin (CPFX), Doxycycline hydrate (DHX), and Tetracycline hydrate (TC). As shown in Fig. 8, the ENF exhibits a notably higher resistance to current than the other antibiotics, demonstrating its specificity. Additionally, the mixture of ENF, DHX, CPFX, and TC reduced the current profile, likely due to the aptamer's preferential interaction with ENF. Overall, these results indicate that the aptasensor can selectively detect ENF.

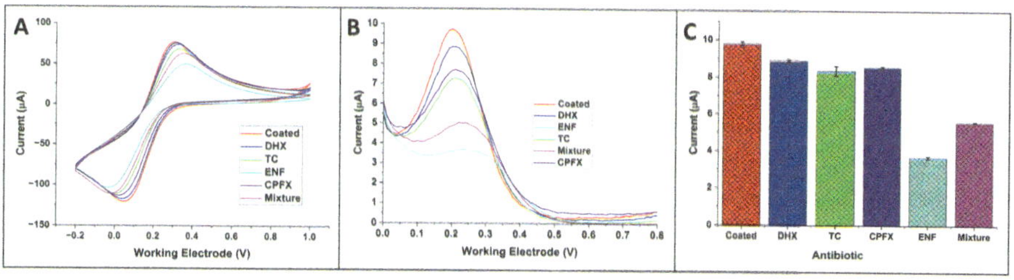

Fig. 8. Specificity analysis of the aptasensor for various antibiotics. DPV (A), CV (B), and bar graph responses for ENF, DHX, CPFX, TC, and ENF + DHX + CPFX + TC (C).

3.5 Detection of Enrofloxacin in Milk

The GCE/rGO-Au/apt electrode was evaluated for detecting enrofloxacin in milk samples using DPV response. The range of enrofloxacin recovery rate was 92.11–115.51%. These results illustrated that the application of the developed GCE/rGO-Au/apt sensor was satisfactory for detecting enrofloxacin. The Recovery and Relative Standard deviation (RSD) % were given in Table 2. The DPV curve of enrofloxacin implies that the presence of other analytes does not interfere with the enrofloxacin (Fig. 9). This shows that the developed GCE/rGO-Au/apt electrode is ideal for detection of enrofloxacin in the presence of other interfering molecules present in the milk sample.

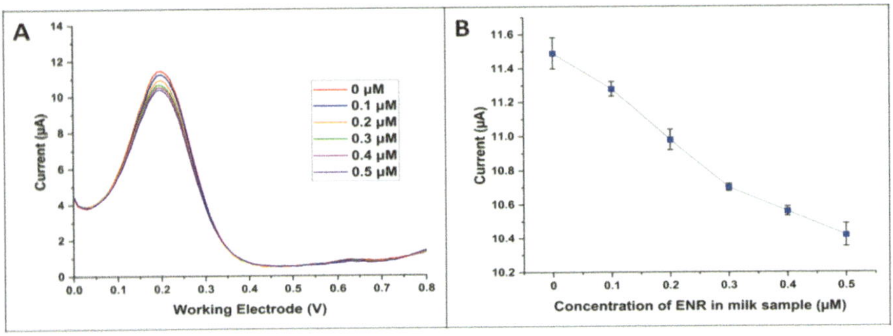

Fig. 9. Real sample detection of enrofloxacin from milk samples at different concentrations.

Table 2. Recovery rate of enrofloxacin in Milk sample.

Sample	Added (μM)	Found (μM)	Recovery %	RSD %
Milk	0.1	0.09	92.11	0.46
	0.2	0.22	111.85	0.66
	0.3	0.35	115.51	0.25
	0.4	0.41	101.98	0.32
	0.5	0.47	93.87	0.80

4 Conclusion

In this study we successfully developed a label-free electrochemical aptasensor utilizing an aptamer as the bioreceptor element to detect Enrofloxacin with high sensitivity, selectivity, stability, and repeatability. The rGO-AuNP nanocomposite enhances the sensor's electrochemical properties, providing a larger surface area, improved electron transfer, and higher conductivity. These characteristics facilitate the aptamer immobilization and provide a robust platform for the detection of low concentrations of Enrofloxacin with high accuracy. The rGO-Au nanocomposite-based aptasensor demonstrated exceptional performance in detecting Enrofloxacin through differential pulse voltammetry and cyclic voltammetry, offering a simple, cost-effective, and efficient method for trace-level analysis. Its high selectivity was confirmed by low interference in the presence of other antibiotics (such as DHX and TC), ensuring that Enrofloxacin is the primary analyte detected with high specificity. This aptasensor design, which is both economical and label-free, enables rapid, accurate screening of Enrofloxacin, making it highly suitable for food safety applications. The aptasensor in future could be used to detect Enrofloxacin in various samples such as milk, meat, poultry, eggs, pharmaceuticals, and drinking water underscores its utility in monitoring contamination and ensuring public health. The findings present the aptasensor as a robust platform for future deployment in food quality control and environmental safety, with potential for broader applications in therapeutic monitoring and diagnosis.

References

1. Wu, G., Ma, J., Li, S., Guan, J., Jiang, B., Wang, L., et al.: Magnetic copper-based metal organic framework as an effective and recyclable adsorbent for removal of two fluoroquinolone antibiotics from aqueous solutions. J. Colloid Interface Sci. **528**, 360–371 (2018)
2. Shao, H., Zhou, H., Zhang, T., Zhao, X., Jiang, Z., Wang, Q.: Preparation of molecularly imprinted hybrid monoliths for the selective detection of fluoroquinolones in infant formula powders. J. Chromatogr. A **1588**, 33–40 (2019)
3. Hyung, S.-W., Lee, C.-H., Kim, B.: Development of certified reference materials for accurate determination of fluoroquinolone antibiotics in chicken meat. Food Chem. **229**, 472–478 (2017)
4. Wang, Y., Tong, Y., Xu, X., Zhang, L.: Metal-organic framework-derived three-dimensional porous graphitic octahedron carbon cages-encapsulated copper nanoparticles hybrids as highly efficient enrichment material for simultaneous determination of four fluoroquinolones. J. Chromatogr. A **1533**, 1–9 (2018)
5. Tang, Y., Xu, J., Chen, L., Qiu, J., Liu, Y., Ouyang, G.: Rapid in vivo determination of fluoroquinolones in cultured puffer fish (Takifugu obscurus) muscle by solid-phase microextraction coupled with liquid chromatography-tandem mass spectrometry. Talanta **175**, 550–556 (2017)
6. Pugajeva, I., Avsejenko, J., Judjallo, E., Bērziņš, A., Bartkiene, E., Bartkevics, V.: High occurrence rates of enrofloxacin and ciprofloxacin residues in retail poultry meat revealed by an ultra-sensitive mass-spectrometric method, and antimicrobial resistance to fluoroquinolones in Campylobacter spp. Food Addit. Contam. Part A Chem. Anal. Control Expo. Risk Assess. **35**, 1107–1115 (2018)
7. Ha, M.-S., Chung, M.-S., Bae, D.-H.: Simple detection of residual enrofloxacin in meat products using microparticles and biochips. Food Addit. Contam. Part A Chem. Anal. Control Expo. Risk Assess. **33**, 817–823 (2016)
8. Tian, Y., Li, G., Zhang, H., Xu, L., Jiao, A., Chen, F., et al.: Construction of optimized Au@Ag core-shell nanorods for ultralow SERS detection of antibiotic levofloxacin molecules. Opt. Express **26**, 23347–23358 (2018)
9. Aymard, C., Kanso, H., Serrano, M.J., Pagán, R., Noguer, T., Istamboulie, G.: Development of a new dual electrochemical immunosensor for a rapid and sensitive detection of enrofloxacin in meat samples. Food Chem. **370**, 131016 (2022)
10. Liu, X., Ren, J., Su, L., Gao, X., Tang, Y., Ma, T., et al.: Novel hybrid probe based on double recognition of aptamer-molecularly imprinted polymer grafted on upconversion nanoparticles for enrofloxacin sensing. Biosens. Bioelectron. **87**, 203–208 (2017)
11. Tufa, R.A., Pinacho, D.G., Pascual, N., Granados, M., Companyó, R., Marco, M.P.: Development and validation of an enzyme linked immunosorbent assay for fluoroquinolones in animal feeds. Food Control **57**, 195–201 (2015)

12. Lee, H., Lee, S., Kwon, D., Yim, C., Jeon, S.: Microbial respiration-based detection of enrofloxacin in milk using capillary-tube indicators. Sens Actuators B Chem. **244**, 559–564 (2017)
13. Tian, R., Ji, J., Zhou, Y., Du, Y., Bian, X., Zhu, F., et al.: Terminal-conjugated non-aggregated constraints of gold nanoparticles on lateral flow strips for mobile phone readouts of enrofloxacin. Biosens. Bioelectron. **160**, 112218 (2020)
14. Aslam, B., Kousar, N., Javed, I., Raza, A., Ali, A., Khaliq, T., et al.: Determination of enrofloxacin residues in commercial broilers using high performance liquid chromatography. Int. J. Food Prop. **19**, 2463–2470 (2016)
15. Schneider, M.J., Braden, S.E., Reyes-Herrera, I., Donoghue, D.J.: Simultaneous determination of fluoroquinolones and tetracyclines in chicken muscle using HPLC with fluorescence detection. J. Chromatogr. B Analyt. Technol. Biomed. Life Sci. **846**, 8–13 (2007)
16. Xu, H., Chen, L., Sun, L., Sun, X., Du, X., Wang, J., et al.: Microwave-assisted extraction and in situ clean-up for the determination of fluoroquinolone antibiotics in chicken breast muscle by LC-MS/MS. J. Sep. Sci. **34**, 142–149 (2011)
17. Wang, P., Wang, L., Li, C., Li, X., Li, G.: Reliable and rapid detection and quantification of enrofloxacin using a ratiometric SERS aptasensor. Molecules (2022). https://doi.org/10.3390/molecules27248764
18. Pokharkar, V., Bhumkar, D., Suresh, K., Shinde, Y., Gairola, S., Jadhav, S.S.: Gold nanoparticles as a potential carrier for transmucosal vaccine delivery. J. Biomed. Nanotechnol. **7**, 57–59 (2011)
19. Guo, S., Huang, Y., Jiang, Q., Sun, Y., Deng, L., Liang, Z., et al.: Enhanced gene delivery and siRNA silencing by gold nanoparticles coated with charge-reversal polyelectrolyte. ACS Nano **4**, 5505–5511 (2010)
20. Saha, K., Agasti, S.S., Kim, C., Li, X., Rotello, V.M.: Gold nanoparticles in chemical and biological sensing. Chem. Rev. **112**, 2739–2779 (2012)
21. Zayats, M., Baron, R., Popov, I., Willner, I.: Biocatalytic growth of Au nanoparticles: from mechanistic aspects to biosensors design. Nano Lett. **5**, 21–25 (2005)
22. Hall, W.P., Ngatia, S.N., Van Duyne, R.P.: LSPR biosensor signal enhancement using nanoparticle-antibody conjugates. J Phys Chem C Nanomater Interfaces. **115**, 1410–1414 (2011)
23. Pissuwan, D., Cortie, C.H., Valenzuela, S.M., Cortie, M.B.: Functionalised gold nanoparticles for controlling pathogenic bacteria. Trends Biotechnol. **28**, 207–213 (2010)
24. Pivodová, V., Franková, J., Galandáková, A., Ulrichová, J.: AuNPs' cytotoxicity and their effect on wound healing. Nanobiomedicine (Rij). **2**, 7 (2015)
25. Megiel, E.: Surface modification using TEMPO and its derivatives. Adv. Colloid Interface Sci. **250**, 158–184 (2017)
26. Becerril, H.A., Mao, J., Liu, Z., Stoltenberg, R.M., Bao, Z., Chen, Y.: Evaluation of solution-processed reduced graphene oxide films as transparent conductors. ACS Nano **2**, 463–470 (2008)
27. Gao, C., Huang, X.-J.: Voltammetric determination of mercury(II). Trends Analyt Chem. **51**, 1–12 (2013)
28. Cesarino, I., Hümmelgen, I.A.: An additional tool towards overcoming absence of specificity of carbon nanostructure-based electrochemical sensors—application to estriol and estradiol detection and distinction. J. Solid State Electrochem. **19**, 3045–3050 (2015)
29. Shao, Y., Wang, J., Wu, H., Liu, J., Aksay, I.A., Lin, Y.: Graphene based electrochemical sensors and biosensors: a review. Electroanalysis **22**, 1027–1036 (2010)
30. Chen, D., Feng, H., Li, J.: Graphene oxide: preparation, functionalization, and electrochemical applications. Chem. Rev. **112**, 6027–6053 (2012)
31. Ma, P., Duan, N., Ye, H., Xia, Y., Ding, Z., Wang, Z.: Selection, truncation and fluorescence polarization based aptasensor for Weissella viridescens detection. Talanta **246**, 123499 (2022)

32. Xiong, J., Dong, C., Zhang, J., Fang, X., Ni, J., Gan, H., et al.: DNA walker-powered ratiometric SERS cytosensor of circulating tumor cells with single-cell sensitivity. Biosens. Bioelectron. **213**, 114442 (2022)
33. Wang, Q., Liu, L., Yang, X., Wang, K., Chen, N., Zhou, C., et al.: Evaluation of medicine effects on the interaction of myoglobin and its aptamer or antibody using atomic force microscopy. Anal. Chem. **87**, 2242–2248 (2015)
34. Jia, Y., Zhou, G., Wang, X., Zhang, Y., Li, Z., Liu, P., et al.: A metal-organic framework/aptamer system as a fluorescent biosensor for determination of aflatoxin B1 in food samples. Talanta **219**, 121342 (2020)
35. Wu, S., Mao, J., Zhang, Y., Wang, S., Huo, M., Guo, H.: Sensitive electrochemical detection of enrofloxacin in eggs based on carboxylated multi-walled carbon nanotubes-reduced graphene oxide nanocomposites: molecularlyimprintedrecognition versus direct electrocatalytic oxidation. Food Chem. **413**, 135579 (2023)
36. Gallegos-Tabanico, A., Jimenez-Canale, J., Hernandez-Leon, S.G., Burgara-Estrella, A.J., Encinas-Encinas, J.C., Sarabia-Sainz, J.A.: Development of an electrochemical sensor conjugated with molecularly imprinted polymers for the detection of enrofloxacin. Chemosensors (Basel). **10**, 448 (2022)
37. Niu, Z., Wang, S., Luo, R., Wei, P., Li, K., Wang, W., et al.: Electrochemical detection of enrofloxacin in meat using bimetallic organic framework-derived NiCoO@NiO. Anal. Methods **15**, 1441–1451 (2023)
38. Karuppusamy, N., Mariyappan, V., Chen, S.-M., Ramachandran, R.: A novel electrochemical sensor for the detection of enrofloxacin based on a 3D flower-like metal tungstate-incorporated reduced graphene oxide nanocomposite. Nanoscale **14**, 1250–1263 (2022)
39. Zhang, Y., et al.: Aptamer-modified sensitive nanobiosensors for the specific detection of antibiotics. J. Mater. Chem. B **8**(37), 8607–8613 (2020)
40. Kushikawa, R.T., Silva, M.R., Angelo, A.C., Teixeira, M.F.: Construction of an electrochemical sensing platform based on platinum nanoparticles supported on carbon for tetracycline determination. Sens. Actuators B Chem. **228**, 207–213 (2016)
41. del Torno-de, R.L., Alonso-Lomillo, M.A., Domínguez-Renedo, O., Arcos-Martínez, M.J.: Tyrosinase based biosensor for the electrochemical determination of sulfamethoxazole. Sens. Actuators B Chem. **227**, 48–53 (2016)
42. Anh, N.T., Dinh, N.X., Pham, T.N., Le, A.T.: Enhancing the chloramphenicol sensing performance of Cu–MoS 2 nanocomposite-based electrochemical nanosensors: roles of phase composition and copper loading amount. RSC Adv. **11**(49), 30544–30559 (2021)
43. Yuan, Y., Zhang, F., Wang, H., Gao, L., Wang, Z.: A sensor based on Au nanoparticles/carbon nitride/graphene composites for the detection of chloramphenicol and ciprofloxacin. ECS J. Solid State Sci. Technol. **7**(12), M201–M208 (2018)
44. Borowiec, J., Wang, R., Zhu, L., Zhang, J.: Synthesis of nitrogen-doped graphene nanosheets decorated with gold nanoparticles as an improved sensor for electrochemical determination of chloramphenicol. Electrochim. Acta **99**, 138–144 (2013)
45. Wang, F., Zhu, L., Zhang, J.: Electrochemical sensor for levofloxacin based on molecularly imprinted polypyrrole–graphene–gold nanoparticles modified electrode. Sens. Actuators B Chem. **192**, 642–647 (2014)
46. Gan, T., Lv, Z., Liu, N., Shi, Z., Sun, J., Liu, Y.: Electrochemical detection method for chlorotetracycline based on enhancement of yolk–shell structured carbon sphere@ MnO2. J. Electrochem. Soc.Electrochem. Soc. **162**(4), H200 (2015)

Wearable Monitoring of Brain Activations in Response to Visual Stimuli Provided Through a Virtual Reality Headset

Martti Ilvesmäki[1(✉)], Alessandra Dussin[1], Jari Paunonen[1], Atefeh Amin[2], Sadegh Moradi[2], Kristina Mikkonen[1], Hany Ferdinando[1], and Teemu Myllylä[1,2,3]

[1] Research Unit of Health Sciences and Technology, Faculty of Medicine, University of Oulu, Oulu, Finland
martti.ilvesmaki@oulu.fi
[2] Optoelectronics and Measurement Techniques Research Unit, University of Oulu, Oulu, Finland
[3] Medical Research Center (MRC) Oulu, Oulu, Finland

Abstract. Emerging Virtual Reality (VR) technologies are gaining momentum, offering innovative tools for applications in fields such as neuroscience. Coupled with advancements in wearable brain measurement techniques such as electroencephalogram (EEG) and functional near-infrared spectroscopy (fNIRS), these developments enable novel explorations of human brain activation particularly in response to more realistic visual stimuli. We have created a system that integrates EEG and fNIRS within a VR headset, facilitating the monitoring of neuronal activity and cerebral hemodynamics as participants engage in visual tasks.

We examined various visual tasks in connection with the brain's induced responses, focusing on potential differences in brain activity when utilizing a 3D VR headset compared to a conventional 2D display. Differences in cerebral responses were examined utilizing time-domain and frequency-domain analysis methods. Observations from the EEG indicated alterations in theta and beta frequency band powers during both stimuli. Furthermore, both oxygenated hemoglobin and deoxygenated hemoglobin showed different responses during the tasks, suggesting alterations in hemodynamic activity. These preliminary findings offer novel methodological insights into the differences in neural and hemodynamic responses between 2D video and VR stimuli.

Keywords: functional near-infrared spectroscopy · electroencephalography · virtual reality

1 Introduction

Use of virtual reality (VR) provides intriguing applications in numerous fields, such as in neuroscience. VR offers unique opportunities to simulate realistic and controlled environments, making it a valuable tool for studying complex brain responses that may not be easily observed in traditional laboratory settings. The researchers can benefit from

C. T. Lim et al. (Eds.): ICBME 2024, IFMBE Proceedings 132, pp. 106–117, 2025.
https://doi.org/10.1007/978-3-032-04372-6_9

these technologies to study brain responses to particular realistic and immersive settings that VR can offer [1] and it can provide valuable applications to many fields, such as for the treatment of anxiety disorders through VR exposure therapy [2].

To accomplish this aim, the continuous and increased development of advanced non-invasive brain measurement techniques, such as functional near-infrared spectroscopy (fNIRS) and electroencephalogram (EEG), enables the analysis of brain activity [3]. These techniques provide complementary insights into brain function, with EEG offering high temporal resolution to capture rapid neural events, and fNIRS providing spatial information about changes in cerebral blood flow and oxygenation. Both methods are non-invasive, portable, and increasingly accessible, making them suitable for integration with VR setups.

VR together with EEG and fNIRS has already been used for brain monitoring in various fields, such as in rehabilitation [4]. For instance, VR-based rehabilitation systems combined with EEG and fNIRS allow clinicians to monitor real-time neural and hemodynamic changes during motor or cognitive tasks, enabling personalized treatment strategies. In these studies, VR is predominantly utilized in conjunction with EEG. Therefore, the inclusion of fNIRS in the multimodal monitoring setup provides more information about the brain activity during the task compared to using a single neuroimaging modality [5]. Additionally, fNIRS and EEG together improve the spatial and temporal resolution of the signal, while the near-infrared light of fNIRS does not interfere with EEG's electrical measurements [3].

There is growing evidence indicating that VR environments provide a greater sense of presence in adults when compared to conventional 2D screen as measured with subjective scales, electroencephalography (EEG) and functional magnetic resonance imaging (fMRI) [6–8]. Combination of VR headset and brain monitoring has been studied with various VR environments and tasks, such as navigation task [6, 7] and meditation [9]. These tasks leverage VR's capacity to create dynamic, interactive, and immersive experiences, which are challenging to replicate in conventional 2D setups. For instance, navigation tasks in VR can simulate realistic spatial environments that engage brain regions involved in spatial memory and orientation. The studies commonly investigate comparison between more immersive and less-immersive systems, where the same tasks and environments are applied with conventional 2D screen and VR headset, and neural responses are recorded [6, 7, 9]. These comparisons help to illustrate the neural mechanisms underlying immersion and presence by examining differences in brain activity patterns across varying levels of immersion.

Distinct brain activation patterns have been detected with VR and brain monitoring experiments, especially related to the subjective experience of presence [6–9]. In addition to fMRI and EEG, functional near infrared spectroscopy (fNIRS) has been combined with VR with a growing interest [10]. This integration is gaining traction due to the portability of fNIRS, which allows for real-time monitoring of cerebral hemodynamics in naturalistic and interactive VR settings.

We have created a system that integrates EEG and fNIRS within a VR headset, facilitating the monitoring of neuronal activity and cerebral hemodynamics as participants engage in watching a short video with VR headset and 2D screen. To the best of our knowledge, this is the first study to investigate effect of 2D and VR headset with combined fNIRS and EEG setup.

2 Methods and Data

Total of 11 subjects were measured, each subject going through the same protocol: a 5-min baseline, followed by a 2.5-min video of ballet dancing with VR headset and bouncing ball with a 2D display (Fig. 1). VR headset and 2D measurements were done in separate sessions. After data quality verification, five (age: 26 ± 5.1 years, 2 females) of the measurement data were analyzed for the 2D group and four (age: 26.3 ± 5.8 years, 2 females) for the VR headset group. fNIRS data quality was verified by visualizing raw signals time-series and magnitude spectra. If expected peak in magnitude spectrum around 1 Hz, contributing to cardiac activity, was not visible, or large movement artefacts were present, the measurement was discarded from further analysis.

Our fNIRS system [11] utilized wavelengths of 690 nm and 830 nm, which are sensitive to deoxygenated hemoglobin (HbR) and oxygenated hemoglobin (HbO) respectively [12]. Both fNIRS and EEG sensors were placed on the Fp2 region of the International 10–20 system for EEG. The frontal lobe was chosen for this study because 1) it is free of hair, reducing interference in fNIRS measurements, and 2) frontal theta activity is associated with cognitive control [13] and modulation of presence has been associated with dorsolateral prefrontal cortex [7]. The data were sampled at a sampling frequency of 250 Hz.

Data processing and analysis was done with MATLAB R2023b. Modified beer lambert law (MBBL, Eq. 1) was applied to compute HbO and HbR relative concentrations [14]

$$\Delta OD = -\log\left(\frac{I}{I_0}\right) = \varepsilon \Delta CLB \tag{1}$$

where I is the intensity of received light, I_0 is the incident light, ε is the extinction coefficient at wavelength λ, C is the chromophore concentration, L is the optical path length, OD is the optical density and B is differential path length factor (DPF). By using two or more wavelengths around the isobestic point of the chromophore of interest, we can calculate the changes in concentrations of different hemoglobin species according to the following equation (Eq. 2):

$$\Delta OD^{\lambda} = \left(\varepsilon_{HbO}^{\lambda} \Delta C_{HbO} + \varepsilon_{HbR}^{\lambda} \Delta C_{HbR}\right) B^{\lambda} \tag{2}$$

The intensity of the transmitted light I_0 was not known, thus we normalized the recorded light intensity I with its corresponding mean. The optical path length, or source detector separation distance, was set to 3 cm to ensure sufficient light penetration to the cortical layer [15]. The DPF was set to 5.93 in our computations [16].

We analyzed fNIRS signal within very low-frequency (VLF, 0.008–0.1 Hz) bandwidth, while EEG analysis focused between 0.5 and 25 Hz bandwidth, where especially two EEG bandwidths were in the focus of our interest: 1) EEG theta, which is linked to concentration, attention and memory [9, 13], and 2) EEG beta, which is associated with active concentration [17], executive function [18] and suppression of undesirable emotions [19]. When required by analysis method, the signals were filtered with ideal zero-phase filters with corresponding cutoff frequencies for each signal.

Our analysis methods focus on four approaches, 1) analysis of standard deviations, 2) analysis of hemodynamic and relative power responses, 3) spectral entropy (SE) analysis and 4) magnitude spectrum analysis. Standard deviation analysis was applied to time-domain signals to assess how values from different stimulation methods deviated from their corresponding baselines. This analysis reflects the oscillation amplitude resulting from neuronal activity, including both electrical activity and the associated hemodynamic

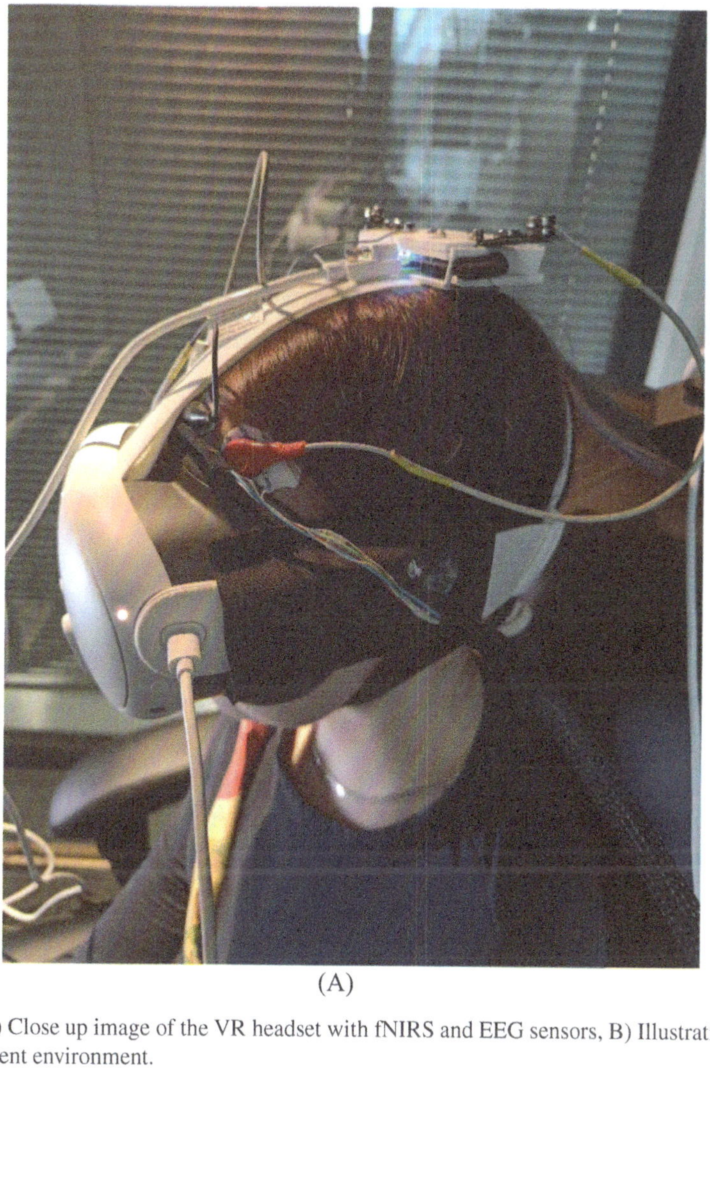

(A)

Fig. 1. A) Close up image of the VR headset with fNIRS and EEG sensors, B) Illustration of the measurement environment.

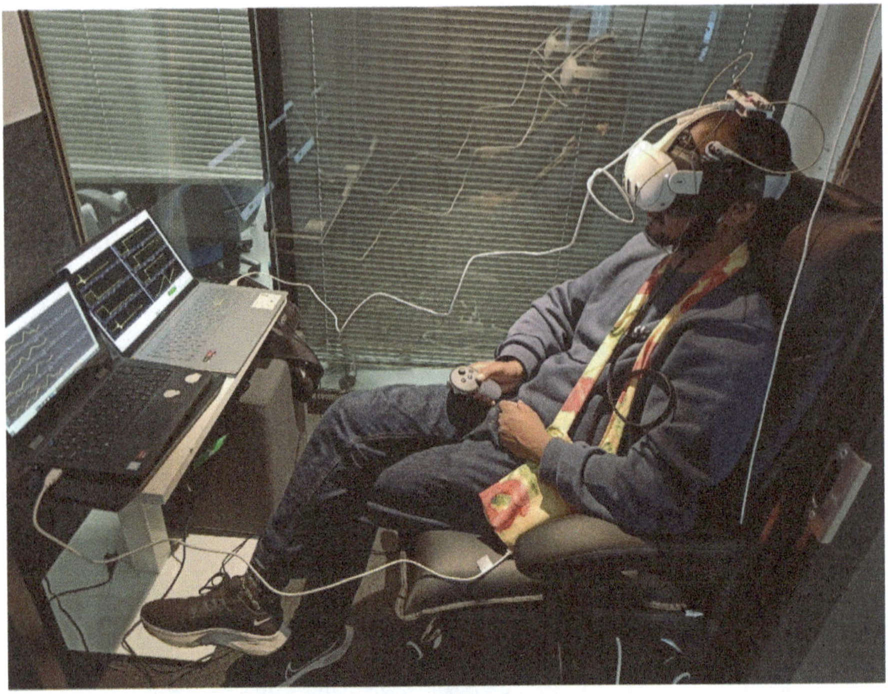

(B)

Fig. 1. (*continued*)

response. We compared changes in standard deviation between 2D and VR stimulations relative to the baseline.

From fNIRS, we observed hemodynamic response, i.e., HbO and HbR responses, from the whole measurement time course. We expected to see changes that can be visually observed between baseline and stimulation. Relative power analysis was applied to EEG signals. It is a common method to analyze how the band powers compare to each other. In this analysis, we used the following bandwidths: delta (0.5–4 Hz), theta (4–8 Hz), alpha (8–12 Hz) and beta (12–25 Hz). The power was calculated using spectrogram method at window size of 30 s with overlap of 29 s.

Spectral entropy (SE) quantifies the complexity of the signal in frequency-domain. Higher complexity in oscillations results in higher SE score [20]. When certain oscillations dominate the signal in comparison to others, then the score becomes lower. We additionally observed the magnitude spectrum from both fNIRS and EEG. For this analysis, we only observed magnitude spectrum within 0–5 Hz for fNIRS and 0–25 for EEG.

3 Results

The differences between 2D and VR headset standard deviations show the largest differences between HbO and EEG theta as illustrated in Fig. 2. It can be observed that the direction of the change is same with each signal. The absolute standard deviation change of HbO is largest with 2D measurement and for EEG theta with VR headset.

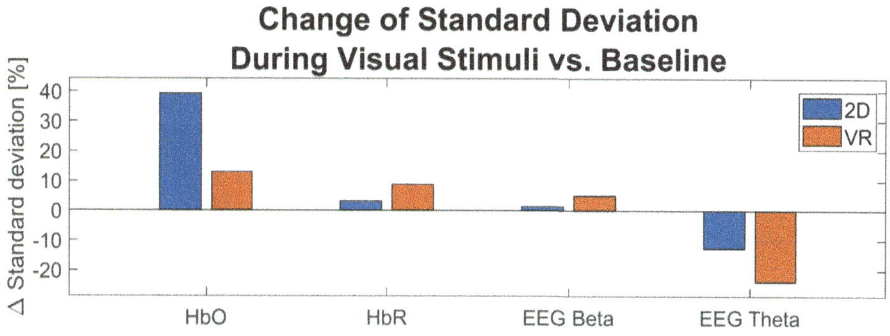

Fig. 2. Differences in standard deviations for fNIRS HbO and HbR relative concentrations and EEG beta and theta relative powers.

Hemodynamic and EEG theta/beta relative powers show rather similar responses in time domain (Fig. 3). However, as shown in Fig. 4, HbR and EEG show increased SE with the 2D screen as compared to the VR headset. Furthermore, magnitude spectrum analysis (Fig. 5) shows 2D HbR contains greater frequency power than baseline in ~ 0.008–3 Hz and VR headset in 0.008–2 Hz approximately. HbO (both VR and 2D) and EEG with the 2D screen show rather similar responses between video stimuli and baseline, while VR EEG shows clear magnitude spectra differences.

4 Discussion

Our findings illustrate changes in neural and hemodynamic responses when providing visual stimuli with 2D and VR headset. These recorded dynamics may stem from various underlying factors, warranting further investigation. The detected alternations could reflect differences in attention, presence [5] and concentration. The more contained environment of VR headset may decrease the effort required for holding attention and decrease attentional fatigue [21] due to increased experience of presence [7]. On the other hand, the observed differences could also be related to increased cognitive load [22].

Standard deviation changes, as shown in Fig. 2, may indicate that during 2D stimulation, the brain engaged in more intensive neural processing, such that it required more oxygenated blood to support neuronal activation. 2D stimulation was not as realistic as the video shown with VR headset, which may induce higher cognitive load [23]. Consequently, the brain needs more energy from the oxygenated blood to process the

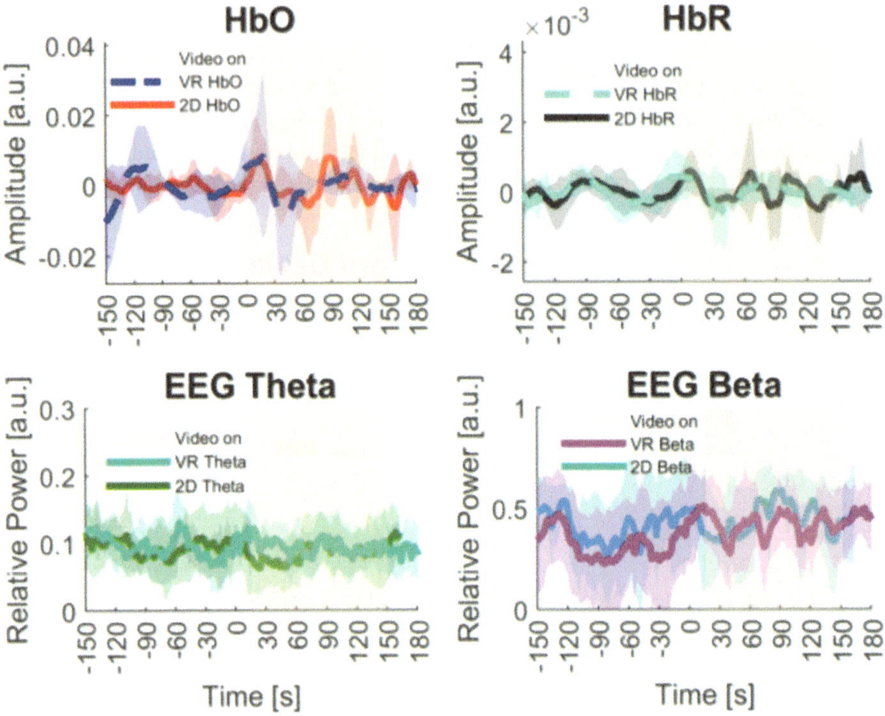

Fig. 3. fNIRS hemodynamic and EEG relative power responses to 2D display and VR headset stimuli.

information. Standard deviation changes in the theta band for 2D were also lower than that for VR. It may suggest higher cognitive load to process less immersive stimulation.

EEG beta activity has been associated with visual attention [24]. As shown in Fig. 5, our result supports this hypothesis, as with VR headset the beta frequency band power increases when compared to baseline. Increased theta power is associated with enhanced cognitive processing, while decreased power corresponds to diminished task engagement [6]. Moreover in the context of a VR environment, higher power of frontal theta has been observed during a VR experience compared to a 2D experience during spatial navigation task [6]. Our result shows contrary result (Fig. 5) as VR theta power is higher in baseline than during VR headset measurement. This difference may be attributed to the distinct cognitive demands of our passive viewing paradigm, which might have involved different cognitive processes compared to the active tasks. Additionally, differences in VR environment design and individual variability could have contributed to the observed discrepancy. Future research could benefit from investigating the role of VR immersion and individual variability to clarify their influence on theta oscillations in VR.

In the context of fNIRS and VR, it has been shown that the average concentration of HbO increases with the level of difficulty of the attention task [25]. Although our findings do not support the finding directly, as shown by fairly similar average hemodynamic responses with VR headset and 2D screen, the data highlights other distinctions. Lower

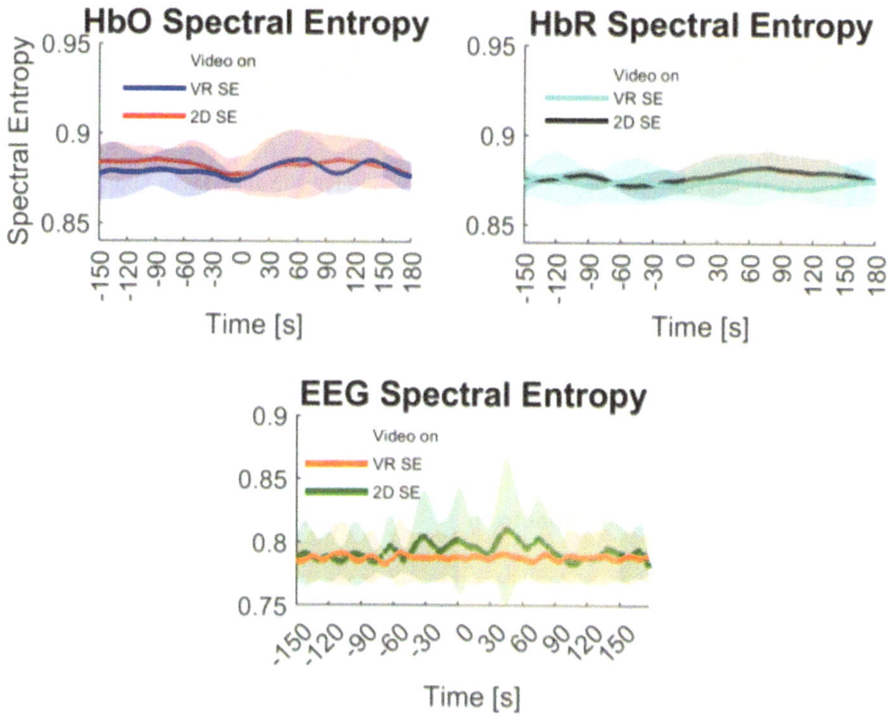

Fig. 4. Spectral entropy time domain responses with 2D display and VR headset.

change in standard deviation of HbO can be observed as measured with VR headset (Fig. 2) when compared to the change measured with 2D screen protocol. Further, decreased overall spectral power with HbR was recorded with VR headset (Fig. 5). In the case of HbR, the average SE is lower with VR headset than with 2D screen (Fig. 4).

Although our results are promising, there are several limitations that need to be addressed. Firstly, larger sample size is needed to validate our findings. A small sample size can limit the generalizability of the results, as individual differences, such as variations in baseline neural activity, cognitive ability, or familiarity with VR, may introduce biases that affect the outcome. With a larger sample, the variability between participants can be better accounted for, leading to more reliable estimates of the neural and hemodynamic responses associated with different stimuli. Moreover, a larger cohort would allow for more robust statistical analyses, helping to distinguish subtle effects and ensuring that the observed results are not due to chance or specific to a small group of participants. In turn, this would strengthen the external validity of the findings, making them more applicable to a wider population and enabling more confident conclusions about the broader implications of the study. Inter-subject variability may also be influenced by a range of physiological and behavioral factors, including habitual alcohol and caffeine intake, perceived psychological stress, and physical activity levels. These variables are known to modulate heart rate, heart rate variability and autonomic nervous system activity [26–29], which, in turn, can significantly impact observed hemodynamic and neural responses. Despite the small sample size and the aforementioned uncontrolled factors, we mitigated inter-subject variability by validating the data quality of the recorded data and by instructing the subjects to minimize unnecessary head movements during the recordings.

Secondly, inclusion of stimuli that induces a strong effect on the experience of presence or attention would be of great interest, as well as the analysis of changes it may induce on neurovascular coupling. In our study setup, we used regular videos without a specific rationale for selection, thus not including a cognitive task or pre-designed moment in video that would induce emotional reaction. The strong visual stimuli caused response could be achieved by utilizing a video that promotes a strong sense of presence, such as by using navigation task [6, 7], or by inducing strong emotional response, such as with virtual roller coaster [30]. Future work may benefit from investigating how different VR stimuli, known to elicit heightened neural responses, impact both hemodynamic and

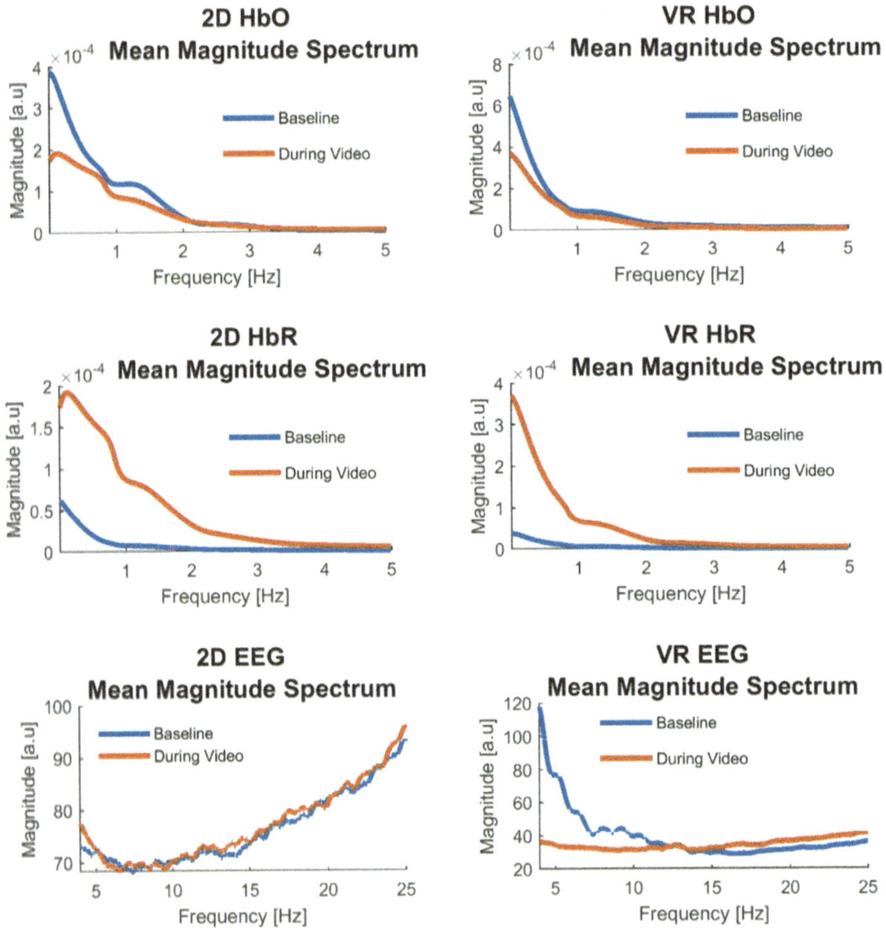

Fig. 5. Magnitude spectra differences between baseline and visual stimuli.

electrical activity in the brain. For example, tasks that demand more complex cognitive processing or emotional engagement could reveal further insights into how immersion influences neural dynamics in VR settings.

Our study presents a novel combination of measuring both EEG and fNIRS when conducting a video stimuli neural response comparison study by using VR headset and 2D screen. Our methodological findings provide valuable insights for the analysis of neural responses, as measured by fNIRS and EEG, in the context of comparing 2D and VR environments.

References

1. Kim, G., Jeon, J., Biocca, F.: M.I.N.D. Brain sensor caps: coupling precise brain imaging to virtual reality head-mounted displays. In: Schmorrow, D.D., Fidopiastis, C.M. (eds.) Augmented Cognition: Intelligent Technologies. pp. 120–130. Springer International Publishing, Cham (2018). https://doi.org/10.1007/978-3-319-91470-1_11
2. Powers, M.B., Emmelkamp, P.M.G.: Virtual reality exposure therapy for anxiety disorders: a meta-analysis. J. Anxiety Disord. **22**, 561–569 (2008). https://doi.org/10.1016/j.janxdis.2007.04.006
3. Shin, J., von Lühmann, A., Kim, D.-W., Mehnert, J., Hwang, H.-J., Müller, K.-R.: Simultaneous acquisition of EEG and NIRS during cognitive tasks for an open access dataset. Sci. Data. **5**, 180003 (2018). https://doi.org/10.1038/sdata.2018.3
4. Choi, J.W., et al.: Neural applications using immersive virtual reality: a review on EEG studies. IEEE Trans. Neural Syst. Rehabil. Eng. **31**, 1645–1658 (2023). https://doi.org/10.1109/TNSRE.2023.3254551
5. Han, C.-H., Müller, K.-R., Hwang, H.-J.: Enhanced performance of a brain switch by simultaneous use of EEG and NIRS data for asynchronous brain-computer interface. IEEE Trans. Neural Syst. Rehabil. Eng. **28**, 2102–2112 (2020). https://doi.org/10.1109/TNSRE.2020.3017167
6. Slobounov, S.M., Ray, W., Johnson, B., Slobounov, E., Newell, K.M.: Modulation of cortical activity in 2D versus 3D virtual reality environments: an EEG study. Int. J. Psychophysiol. **95**, 254–260 (2015). https://doi.org/10.1016/j.ijpsycho.2014.11.003
7. Kober, S.E., Kurzmann, J., Neuper, C.: Cortical correlate of spatial presence in 2D and 3D interactive virtual reality: an EEG study. Int. J. Psychophysiol. **83**, 365–374 (2012). https://doi.org/10.1016/j.ijpsycho.2011.12.003
8. Baumgartner, T., Speck, D., Wettstein, D., Masnari, O., Beeli, G., Jäncke, L.: Feeling present in arousing virtual reality worlds: prefrontal brain regions differentially orchestrate presence experience in adults and children. Front. Hum. Neurosci. (2008). https://doi.org/10.3389/neuro.09.008.2008
9. Kosunen, I., Salminen, M., Järvelä, S., Ruonala, A., Ravaja, N., Jacucci, G.: RelaWorld: Neuroadaptive and Immersive Virtual Reality Meditation System. In: Proceedings of the 21st International Conference on Intelligent User Interfaces, vol. 2, pp. 208–217. Association for Computing Machinery, New York, NY, USA (2016). https://doi.org/10.1145/2856767.2856796
10. Peng, K., Moussavi, Z., Karunakaran, K.D., Borsook, D., Lesage, F., Nguyen, D.K.: IVR-fNIRS: studying brain functions in a fully immersive virtual environment. NPh. **11**, 020601 (2024). https://doi.org/10.1117/1.NPh.11.2.020601
11. Myllylä, T., Korhonen, V., Surażyński, Ł, Zienkiewicz, A., Sorvoja, H., Myllylä, R.: Measurement of cerebral blood flow and metabolism using high power light-emitting diodes. Measurement **58**, 387–393 (2014). https://doi.org/10.1016/j.measurement.2014.08.027

12. Cope, M.: The Application of Near Infrared Spectroscopy to Non Invasive Monitoring of Cerebral Oxygenation in the Newborn Infant. University College London (1991)
13. Cavanagh, J.F., Frank, M.J.: Frontal theta as a mechanism for cognitive control. Trends Cogn. Sci. **18**, 414–421 (2014). https://doi.org/10.1016/j.tics.2014.04.012
14. Boas, D.A., Gaudette, T., Strangman, G., Cheng, X., Marota, J.J.A., Mandeville, J.B.: The accuracy of near infrared spectroscopy and imaging during focal changes in cerebral hemodynamics. Neuroimage **13**, 76–90 (2001). https://doi.org/10.1006/nimg.2000.0674
15. Strangman, G.E., Li, Z., Zhang, Q.: Depth sensitivity and source-detector separations for near infrared spectroscopy based on the Colin27 brain template. PLoS ONE **8**, e66319 (2013). https://doi.org/10.1371/journal.pone.0066319
16. van der Zee, P., Cope, M., Arridge, S.R., Essenpreis, M., Potter, L.A., Edwards, A.D., Wyatt, J.S., McCormick, D.C., Roth, S.C., Reynolds, E.O.R., Delpy, D.T.: Experimentally measured optical pathlengths for the adult head, calf and forearm and the head of the newborn infant as a function of inter optode spacing. In: Goldstick, T.K., McCabe, M., and Maguire, D.J. (eds.) Oxygen Transport to Tissue XIII. pp. 143–153. Springer US, Boston, MA (1992). https://doi.org/10.1007/978-1-4615-3404-4_17
17. Benchenane, K., Tiesinga, P.H., Battaglia, F.P.: Oscillations in the prefrontal cortex: a gateway to memory and attention. Curr. Opin. Neurobiol. **21**, 475–485 (2011). https://doi.org/10.1016/j.conb.2011.01.004
18. Fernández, T., et al.: EEG activation patterns during the performance of tasks involving different components of mental calculation. Electroencephalogr. Clin. Neurophysiol. **94**, 175–182 (1995). https://doi.org/10.1016/0013-4694(94)00262-J
19. Shetty, S.J., Shetty, S., Shettigar, D., Pagilla, V., Maiya, G.A.: Effect of transcranial photobiomodulation on electrophysiological activity of brain in healthy individuals: a scoping review. J. Clin. Neurosci. **117**, 156–167 (2023). https://doi.org/10.1016/j.jocn.2023.09.029
20. Ferdinando, H., Moradi, S., Korhonen, V., Helakari, H., Kiviniemi, V., Myllylä, T.: Spectral entropy provides separation between Alzheimer's disease patients and controls: a study of fNIRS. Eur. Phys. J. Spec. Top. **232**, 655–662 (2022). https://doi.org/10.1140/epjs/s11734-022-00753-w
21. Gao, T., Zhang, T., Zhu, L., Gao, Y., Qiu, L.: Exploring psychophysiological restoration and individual preference in the different environments based on virtual reality. Int. J. Environ. Res. Public Health **16**, 3102 (2019). https://doi.org/10.3390/ijerph16173102
22. Makransky, G., Terkildsen, T.S., Mayer, R.E.: Adding immersive virtual reality to a science lab simulation causes more presence but less learning. Learn. Instr. **60**, 225–236 (2019). https://doi.org/10.1016/j.learninstruc.2017.12.007
23. Kisker, J., Johnsdorf, M., Sagehorn, M., Hofmann, T., Gruber, T., Schöne, B.: Visual information processing of 2D, virtual 3D and real-world objects marked by theta band responses: visuospatial processing and cognitive load as a function of modality. Eur. J. Neurosci. **61**, e16634 (2025). https://doi.org/10.1111/ejn.16634
24. Ahirwal, M.K., Londhe, N.: Power spectrum analysis of EEG signals for estimating visual attention. IJCA **42**, 34–40 (2012). https://doi.org/10.5120/5769-7993
25. Jahani, S., Berivanlou, N.H., Rahimpour, A., Setarehdan, S.K.: Attention level quantification during a modified stroop color word experiment: an fNIRS based study. In: 2015 22nd Iranian Conference on Biomedical Engineering (ICBME). pp. 99–103 (2015). https://doi.org/10.1109/ICBME.2015.7404124
26. Kim, H.-G., Cheon, E.-J., Bai, D.-S., Lee, Y.H., Koo, B.-H.: Stress and heart rate variability: a meta-analysis and review of the literature. Psychiatry Investig. **15**, 235–245 (2018). https://doi.org/10.30773/pi.2017.08.17
27. Fu, Q., Levine, B.D.: Chapter 13 - Exercise and the autonomic nervous system. In: Buijs, R.M. and Swaab, D.F. (eds.) Handbook of Clinical Neurology. pp. 147–160. Elsevier (2013). https://doi.org/10.1016/B978-0-444-53491-0.00013-4

28. Ralevski, E., Petrakis, I., Altemus, M.: Heart rate variability in alcohol use: a review. Pharmacol. Biochem. Behav. **176**, 83–92 (2019). https://doi.org/10.1016/j.pbb.2018.12.003
29. Koenig, J., et al.: Impact of caffeine on heart rate variability: a systematic review. J. Caffeine Res. **3**, 22–37 (2013). https://doi.org/10.1089/jcr.2013.0009
30. Davis, S., Nesbitt, K., Nalivaiko, E.: Comparing the onset of cybersickness using the Oculus Rift and two virtual roller coasters. In: Proceedings of the 11th Australasian Conference on Interactive Entertainment (IE 2015) (2015). https://crpit.scem.westernsydney.edu.au/confpapers/CRPITV167Davis.pdf

Nuclear Medicine Imaging in Parkinson's Disease - Comma or No Comma? - That is the Question

Subapriya Suppiah[1,2,3](✉), Nur Azra Md Zin[3], Ahmad Danial Ahmad Shahrir[3], Mohd Khalil Saleh[3], Mohd Nawawi Abdullah[3], and Vengkatha Priya Seriramulu[2]

[1] Pusat Pengimejan Diagnostik Nuklear, Universiti Putra Malaysia, 43400 Serdang, Malaysia
subapriya@upm.edu.my

[2] Department of Radiology, Faculty of Medicine and Health Sciences, Universiti Putra Malaysia, 43400 Serdang, Malaysia

[3] Unit Pengimejan Nuklear, Hospital Sultan Abdul Aziz Shah, Universiti Putra Malaysia, 43400 Serdang, Malaysia

Abstract. Parkinson's disease and Parkinson's plus syndromes are a group of movement disorders involving the striatonigral pathway. MRI can determine the condition to a certain extent but with limited specificity, particularly in the early stages. Nuclear medicine imaging helps provide improved sensitivity and specificity particularly using ^{18}F-DOPA PET-CT or Iodine ioflupane ^{123}I FP-CIT (DaTscan) SPECT-CT imaging. Nevertheless, the emerging role of ^{99m}Technetium (^{99m}Tc)-based radiotracers such as ^{99m}Tc-TRODAT-1 SPECT-CT imaging provides a good alternative for centers that have limited access to a PET-CT scanners and non ^{99m}Tc-based radiotracers. This write-up is a didactic sharing of experience in nuclear medicine imaging of Parkinson's disease at a Malaysian tertiary, teaching hospital.

Keywords: Parkinson's disease · movement disorder · dementia · diagnostic imaging · dopamine receptor

1 Introduction

Parkinson's disease (PD) and Parkinson's dementia are phenotypically different but with similar underlying pathology of synucleinopathy related to degeneration of the nigrostriatal pathways and ultimately dopamine transmitter abnormalities. The clinical manifestations of Parkinsonism, which is the broader term for the clinical syndrome, include rigidity and muscle stiffness, tremors, bradykinesia, and poor coordination leading to profound gait impairment [1]. It can also manifest as impairment of cognitive functions such as memory loss at the other end of the spectrum of Parkinson's dementia. Additionally, there exist a kaleidoscope of conditions that mimic symptoms of PD, such as essential tremor, drug-induced Parkinsonism (DIP), multiple system atrophy (MSA), progressive supranuclear palsy (PSP), corticobasal degeneration (CBD), and dementia with Lewy bodies (DLB), grouped as the Parkinsonian syndromes, that can be

C. T. Lim et al. (Eds.): ICBME 2024, IFMBE Proceedings 132, pp. 118–123, 2025.
https://doi.org/10.1007/978-3-032-04372-6_10

challenging to differentiate clinically from idiopathic PD [1]. At Hospital Sultan Abdul Aziz Shah (HSAAS), Universiti Putra Malaysia, a tertiary teaching hospital in Malaysia, magnetic resonance imaging (MRI) of the brain is usually a first line imaging to aid in the management of patients presenting with neurocognitive and neuromotor symptoms. From a survey conducted among 112 patients from the geriatric/ memory clinic HSAAS who were referred to perform MRI brain scans for presenting with indeterminate features of either overlapping tauopathy or alpha-synucleinopathies (SNP) features or inconclusive clinical features, during the period of 2022 to 2024, approximately one fifth of the patients were diagnosed to have PD based on the MRI features (Fig. 1).

2 Diagnostic Imaging Using Magnetic Resonance Imaging

Magnetic resonance imaging (MRI) uses multiple sequences such as T1 weighted imaging for excellent anatomical localization and detection of regional atrophy, T2 weighted imaging and FLAIR sequences to evaluate white matter ischemia and cerebral infarction, diffusion weighted imaging (DWI) to assess for acute infarction, and susceptibility weighted imaging (SWI), which normally should demonstrate high signal intensity of nigrosome 1 at the posterior third of the midbrain (dorsolateral nigral hyperintensity) that is bordered by linear hypointense signals of the substantia nigra giving the appearance of the tail of a swallow (Fig. 2a).Conversely, in PD there is absence of the hyperintense signal in the substantia nigra, giving the appearance of loss of the 'swallow tail sign' (STS) (Fig. 2b). Identifying the STS for PD or atypical Parkinsonian syndrome compared to healthy controls on MRI is variable due to lack of standardization of protocols, leading to a wide range of diagnostic accuracy of 47.8–98.5% [2]. Additionally, the loss of the STS has a diagnostic accuracy of 76–90% for differentiating from controls DLB from other dementias [2]. However, the STS is unable to be a reliable biomarker to differentiate PD from atypical Parkinsonian syndrome. The wide range of diagnostic accuracy of the STS in detecting PD can be attributed to several factors including hardware factor caused by the differences in the magnetic field strengths of the MRI scanners, such as the usage of 1.5 T vs. 3.0 T machines, whereby the usage of high-resolution images from higher field strengths can improve the detection of subtle changes in the substantia nigra; patient-selection bias factor caused by different severity or stages of the disease and existence of co-morbidities that can obscure the detection of STS; and finally inter-observer variability factor caused by the differences in experience of the reporting radiologist or the differences caused by manual vs. automated assessment methods [3, 4].

3 Diagnostic Imaging Using 99mTc-TRODAT-1 SPECT-CT Imaging

DaTscans can be interpreted by visual assessment of the basal ganglia regions on axial images, whereby reduced uptake in the putamen giving rise to the 'loss of the comma tail sign' is an early biomarker of SNP such as idiopathic Parkinson's disease, DLB, and MSA-Parkinson's type. DATscan has 98% diagnostic accuracy for detection of nigrostriatal cell loss and can be used to differentiate PD from essential tremors, vascular

Parkinsonism (VP) and secondary PD or also known as atypical Parkinsonian syndromes, i.e., drug-induced PD (DIP), MSA-cerebellar type and PSP [5]. In view of the limited availability of pre-synaptic dopamine transporter imaging, using ^{123}I ioflupane, ^{99m}Tc-TRODAT-1 has been increasingly used to help distinguish idiopathic PD from essential tremor and atypical Parkinsonian syndromes. It is especially useful in differentiating PD at the early stage of disease compared to age-matched healthy controls, with a proposed striatal binding to background occipital lobe uptake cutoff ratio of 0.90 [6]. Additionally, symptomatic VP patients with no evidence of 'loss of comma tail sign' on SPECT-CT imaging had no reaction to levodopa treatment, however, the majority of VP patients with decreased striatal binding with presence of 'loss of the comma tail sign' responded well with symptomatic improvement after administration of levodopa [7]. At HSAAS, there have been several initial experiences using ^{99m}Tc-TRODAT-1 SPECT-CT imaging for uncertain cases, with better diagnosed accuracy for detection of the 'loss of comma sign' particularly in cases with indeterminate MRI findings, leading to better diagnostic certainty and improved patient management.

An example of a case with diagnostic uncertainty was an elderly patient with underlying hypertension and history of multifocal cerebral infarctions. The patient presented with right sided weakness and slurred speech and progressive slowness of movement. ^{99m}Tc-TRODAT-1 SPECT-CT imaging was done to differentiate between VP and idiopathic PD. The scan findings visual assessment of the axial scans at the level of the basal ganglia (Fig. 3) showed reduced uptake at bilateral putamen with 'loss of comma tail sign' (left worse than right side); and the semiquantitative assessment revealed right caudate to putamen ratio of 3.10 vs. left caudate to putamen ratio of 2.54, with percentage of asymmetry R: L being 18%. She was diagnosed with presynaptic Parkinsonism and responded well to administration of levodopa.

A study conducted in Malaysia, that retrospectively reviewed ^{99m}Tc-TRODAT-1 scans in a tertiary public hospital, identified 11 out of 16 symptomatic patients to have positive scans and normal scans in 5 patients, leading to 81.3% change in therapeutic management [8]. Comparatively, another study using DaTscans in the UK, identified 40% abnormal scans and 53% normal with the rest being indeterminate among 455 patients: leading to 42% change in management, with approximately 63% patients with abnormal scans starting dopaminergic medication [9].

Overall, a normal ^{99m}Tc-TRODAT-1 scan in patients suspected with DIP carries a reasonably good prognosis of recovery after removal of the causative drug, whereas strongly asymmetrical loss of comma sign may indicate a latent PD and will benefit from removal of any potential dopamine reuptake blocker drugs [10]. A reasonable prognosis of recovery is expected in the DIP with normal SPECT-TRODAT after removing the causative agent. On the other hand, DIP with strongly asymmetric SPECT-TRODAT results may indicate latent PD and symptoms precipitated using dopamine reuptake blocker drugs [10]. Whereas VP patients with an abnormal scan result demonstrate a trend toward a good therapeutic response to dopaminergic agents [10].

4 Conclusion

Dopamine transporter imaging using ^{99m}Tc-TRODAT-1 SPECT-CT is a useful and reliable nuclear medicine diagnostic investigation to differentiate idiopathic PD from essential tremors and other atypical or secondary Parkinsonian syndromes. Improved diagnostic confidence can be achieved with the usage of semiquantitative assessment by the aid of software tools but require comparison with normal ageing changes. An important point for consideration is to include conventional MRI imaging that is necessary as an additional diagnostic tool to correlate with structural changes and certain functional information.

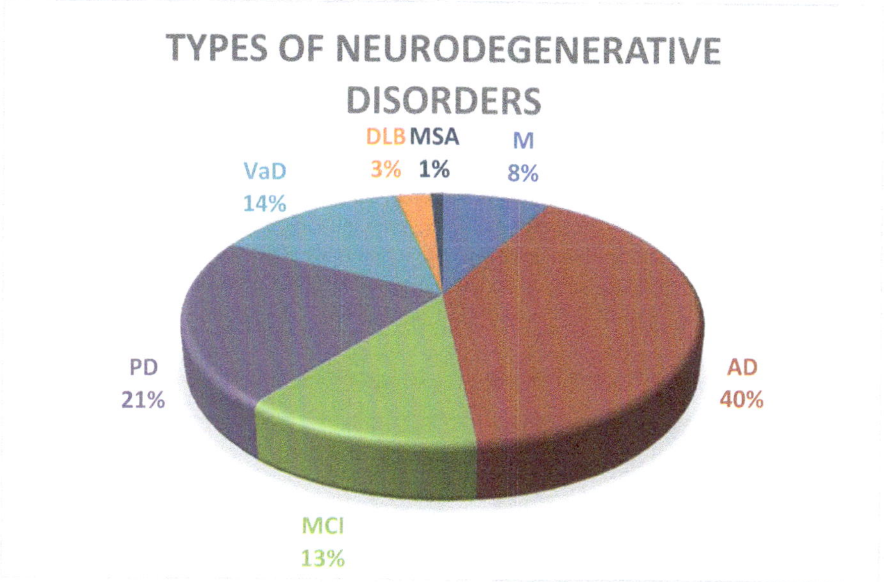

Fig. 1. Distribution of patients with neurodegenerative diseases referred for MRI Brain at Hospital Sultan Abdul Aziz Shah, Universiti Putra Malaysia in 2022–2024 (N = 112) (Abbreviation: M = mixed dementia, AD = Alzheimer's disease, MCI = mild cognitive impairment, PD = Parkinson's disease, VaD = Vascular dementia, DLB = Lewy body dementia, MSA = Multiple system atrophy)

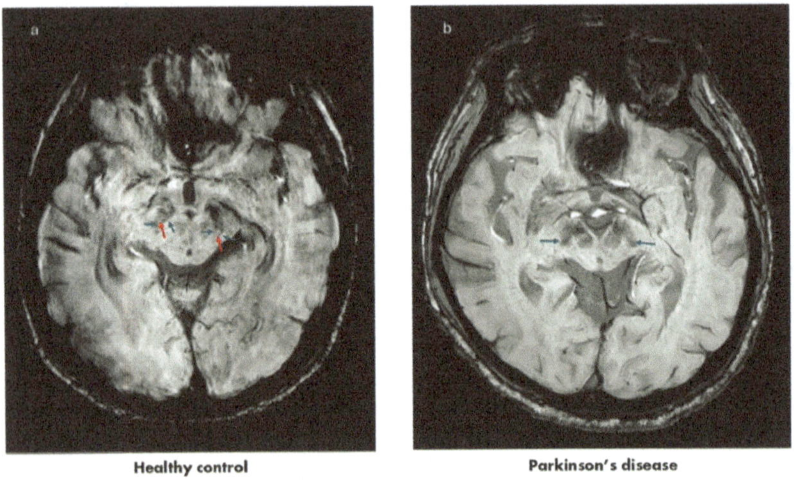

Fig. 2. MRI axial images of susceptibility weighted imaging (SWI) sequence at the level of substantia nigra of the midbrain. (a) high signal intensity (bright linear signal) (red arrow) of nigrosome 1 at the posterior third of the midbrain that is bordered by two dark lines and central bright stripe, i.e., linear hypointense signals (blue arrows) of the substantia nigra giving the appearance of the tail of a swallow. (b) absence of the hyperintense signal (loss of the bright signal between the two dark lines) in the substantia nigra with only the hypointensity (dark lines) seen (blue arrows), giving the appearance of loss of the 'swallow tail sign' (STS).

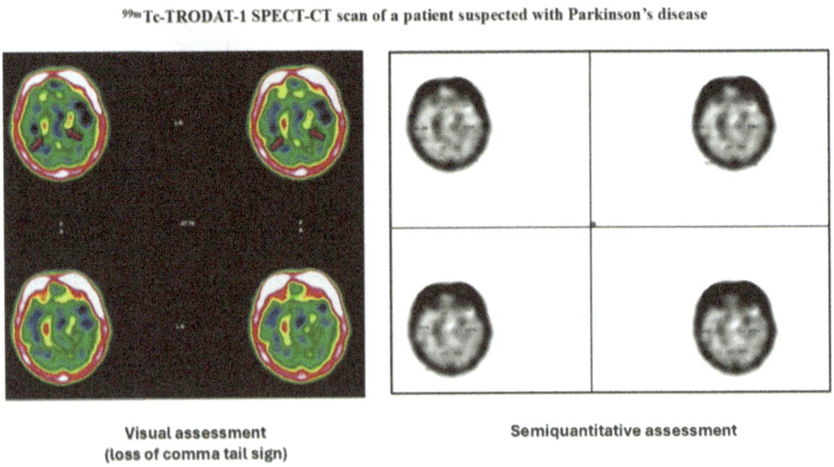

Fig. 3. ^{99m}Tc-TRODAT-1 SPECT-CT of a patient suspected of vascular Parkinsonism. Axial scans at the level of the basal ganglia showed reduced uptake at bilateral putamen with 'loss of comma tail sign' (left worse than right side); and the semiquantitative assessment revealed right caudate to putamen ratio of 3.10 vs. left caudate to putamen ratio of 2.54, with percentage of asymmetry R: L being 18%. She was diagnosed with presynaptic Parkinsonism and responded well to administration of levodopa.

References

1. Murphy, K.T., Lynch, G.S.: Impaired skeletal muscle health in Parkinsonian syndromes: clinical implications, mechanisms and potential treatments. J. Cachexia. Sarcopenia Muscle **14**(5), 1987–2002 (2023)
2. Tseriotis, V.S., Eleftheriadou, K., Mavridis, T., Konstantis, G., Falkenburger, B., Arnaoutoglou, M.: Is the swallow tail sign a useful imaging biomarker in clinical neurology? A systematic review. Movement Disord. Clin. Pract. **12**(2), 134–147 (2025)
3. Zhao, X., Niu, X., You, H., et al.: Signal alteration of Substantia Nigra on 3.0T susceptibility-weighted imaging in Parkinson's disease and vascular parkinsonism. Curr. Med. Sci. **39**, 831–835 (2019)
4. Gupta, R., Kumar, G., Kumar, S., Thakur, B., Tiwari, R., Verma, A.K.: The Swallow Tail Sign of Substantia Nigra: a case-control study to establish its role in diagnosis of parkinson disease on 3T MRI. J. Neurosci. Rural Pract. **13**(2), 181–185 (2022)
5. Vlaar, A.M., van Kroonenburgh, M.J., Kessels, A.G., Weber, W.E.: Meta-analysis of the literature on diagnostic accuracy of SPECT in parkinsonian syndromes. BMC Neurol. **7**(1), 27 (2007)
6. Bor-Seng-Shu, E., et al.: Dopamine transporter imaging using 99mTc-TRODAT-1 SPECT in Parkinson's disease. Med. Sci. Monit. **11**(20), 1413–1418 (2014)
7. Lorberboym, M., Djaldetti, R., Melamed, E., Sadeh, M., Lampl, Y.: 123I-FP-CIT SPECT imaging of dopamine transporters in patients with cerebrovascular disease and clinical diagnosis of vascular parkinsonism. J. Nucl. Med. **45**, 1688–1693 (2004)
8. Kamarulzaman, K., Abd Razak, N., Mawardi, A.S., Amir Hassan, S.Z.: Diagnostic evaluation of Technetium-99 metastable TRODAT-1 single photon emission computed tomography-computed tomography in the differential diagnosis of parkinsonism in Hospital Kuala Lumpur: a preliminary experience. Med. J. Malaysia **79**(6), 690–702 (2024)
9. Tsang, K., Walker, R.: Dopamine transporter single photon emission computed tomography (DaT-SPECT) use in the diagnosis and clinical management of parkinsonism: an 8-year retrospective study. J. Neurol. **270**(5), 2550–2558 (2023)
10. Fabiani, G., Camargo, C.H.F., Filho, R.M., Froehner, G.S., Teive, H.A.G.: Evaluation of brain SPECT with 99mTc-TRODAT-1 in the differential diagnosis of parkinsonism. Parkinson's Dis. 1746540 (2022)

Author Index

C. T. Lim et al. (Eds.): ICBME 2024, IFMBE Proceedings 132, p. 125, 2025.
https://doi.org/10.1007/978-3-032-04372-6

The manufacturer's authorised representative in the EU is Springer Nature Customer Service Centre GmbH, Europaplatz 3, 69115 Heidelberg, Germany. If you have any concerns regarding our products, please contact ProductSafety@springernature.com

Printed and bound by CPI Group (UK) Ltd, Croydon, CR0 4YY
01/10/2026
02225880-0013